Introductory Chemistry: A Workbook

Robert E. Blake, Jr.
Department of Chemistry and Biochemistry
Texas Tech University

Prentice Hall Series in Educational Innovation

PEARSON
Prentice
Hall

Upper Saddle River, NJ 07458

Executive Editor: *Kent Porter Hamann*
Project Manager: *Jacquelyn Howard*
Editor-in-Chief, Science: *John Challice*
Production Editor: *Shari Toron*
Marketing Manager: *Steve Sartori*
Manufacturing Buyer: *Alan Fischer*
AV Editor: *Connie Long*
Art Studio: *Artworks*
Art Director: *Jayne Conte*
Cover Designer: *DaSa Design*
Cover Illustration: *Quade Paul*

© 2005 by Pearson Education, Inc.
Pearson Prentice Hall
Pearson Education, Inc.
Upper Saddle River, New Jersey 07458

Pearson Prentice Hall™ is a trademark of Pearson Education, Inc.

Printed in the United States of America
10 9 8 7 6 5 4 3 2 1

ISBN 0-13-144602-9

Pearson Education LTD., *London*
Pearson Education Australia PTY., Limited, *Sydney*
Pearson Education Singapore, Pte. Ltd.
Pearson Education North Asia Ltd., *Hong Kong*
Pearson Education Canada, Ltd., *Toronto*
Pearson Educación de Mexico, S.A. de C.V.
Pearson Education—Japan, *Tokyo*
Pearson Education Malaysia, Pte. Ltd.

Titles in the Prentice Hall Series in Educational Innovation

CONTENTS

Preface

History

Ten years ago, when I was a graduate student in chemistry, many students lamented that chemistry was one of the hardest subjects they had to take. Despite perfect attendance at lectures, and diligent work and study, many students failed to get the grades they desired. It seemed apparent that some additional instruction or support would be helpful for those students. In chemistry classes, students who formed informal study groups outperformed students who struggled with the material on their own. Because of this, I thought of implementing a systematic program that would provide the benefits of the informal study groups to all students. At that time, I saw a seminar by David Gosser about the Workshop Chemistry model, which had been recently funded by the National Science Foundation. I was so impressed with the design, implementation, and results of the workshops, that I thought it prudent to utilize Workshop Chemistry rather than to figuratively reinvent the wheel by attempting the design of a new, similar program.

When I became an assistant professor of chemistry at Indiana University Purdue University Indianapolis (IUPUI), I wrote a proposal to the National Science Foundation with a consortium of professors who wished to "Adapt and Adopt" Workshop Chemistry to our courses. Thankfully, the NSF funded our proposal (NSF/DUE 9950575), and we were able to implement Workshop Chemistry. The change from traditional discussion sections to the small-group peer-led discussion sections of Workshop Chemistry had a dramatic effect on student performance and reduced the attrition rate in general chemistry at IUPUI by about 40%.

Recently the Workshop Chemistry model has spread into other subjects, so the general method of teaching with peer-led small-group discussion sections has evolved into "Peer-led Team Learning" (PLTL). At Texas Tech University, where I currently teach introductory and general chemistry classes, we decided to use the PLTL model in our introductory chemistry course. Since no PLTL Workbook was available for an introductory chemistry class, the writing of this workbook was undertaken. When a preliminary version was completed, we added peer-led discussion sections to a large lecture class. Over the last two years, several hundred students have used preliminary versions of this workbook in conjunction with that class. More than a hundred peer leaders have directed these students through the workshop activities, and many of these students have provided feedback about the effectiveness of the workbook as a learning tool. This cycle of use, feedback, review, and revision has greatly improved the clarity of the text, the quality of the problems, and the overall usefulness of these materials. I acknowledge the tremendous contributions that the students and peer leaders have made to this workbook, which would certainly not be publishable without their assistance.

Practical Information

The workshops are meant to give students the opportunity to *do* problems and to figure out how to do chemistry. I can show you how to hit a golf ball a thousand times without imparting any ability to hit a golf ball to you. If you want to learn how to hit a golf ball, get a set of clubs. If you want to learn to solve chemistry problems, get a set of problems, a pencil, a calculator, and a periodic table, and then begin to practice.

The workbook is organized into units. Each unit has a brief summary (text) that reviews the chemistry relevant to that unit. This is followed by "Pre-workshop problems," which you should do prior to getting together with your group. This will ensure that everyone is sufficiently well prepared to handle the workshop and not drag the group down. The guts of the unit are found in the "Cooperative Group Problems." This section contains difficult problems, systematic exercises, hands-on activities, and games to assist you in learning chemistry. Finally, a set of multiple-choice "Practice Exam Questions" is provided at the end of each unit to simulate the types of questions your instructor might put on examinations. Even if your instructor favors free-response questions, the practice exam questions will be a good test of what you have learned during the workshop.

The critical components for workshop success are as follows:

1. The PLTL workshops are integral to the course. Small, student-led groups meet for two hours per week. The student leader is someone who has taken the course, has done well in it, and understands how to coach students in problem-solving strategies, rather than someone who wants to do problems for the students in the group.

2. The student leaders are trained in the course content and the facilitation of learning. They must be familiar with strategies to foster collaboration and know how to provide a minimum amount of assistance to enable their students to solve problems. Successful student groups rapidly become cohesive, as they develop their own strategies for working together and solving problems. Successful collaboration reduces their dependence on the student leader.

3. The PLTL materials are of the appropriate level. The materials must be difficult enough to challenge students, but not difficult enough to prevent students from arriving at solutions to the problems.

4. Faculty are involved with the workshops. Faculty should have a hand in arranging the workshop environment, promoting the workshops in class, reviewing the materials, and interacting with the workshop leaders.

5. The environment must be conducive to group discussions and learning. Chairs should be movable, and boards to write on should be available.

6. PLTL has appropriate institutional support. The administration should provide the funding, facilities, and encouragement so that new teaching strategies can be implemented successfully.

This method of learning has been proven to be successful in many disciplines and institutions. Because it is probably different from what you are used to, you might take a little while to adjust to it. Learning to work effectively in teams and to being able to solve problems on your own are the skills your future employer most wants you to have. These are the skills I hope that my workbook and your group will help you to develop. Good luck!

Robert E. Blake, Jr.
bob.blake@ttu.edu

Acknowledgments

The author would like to thank the following people who reviewed *Introductory Chemistry: A Workbook* during its development:

Sean Birke, Jefferson College

Bryan E. Breyfogle, Southwest Missouri State University

Donna Friedman, St. Louis Community College- Florissant Valley

Donna K. Howell, Angelo State University

T.G. Jackson, University of South Alabama

Kirk Kawagoe, Fresno City College

Greg J. Maloney, Northland Pioneer College

C. Michael McCallum, University of the Pacific

John Oakes, Grossmont College

Gita Perkins, Estrella Mountain Community College

Rill A. Reuter, Winona State University

The author would also like to add a special thanks to Michael A. Hauser at St. Louis Community College-Meramec, who reviewed the manuscript for accuracy prior to publication.

R.E.B.

Unit 1: Mathematics as a Tool for Science and Scientists

The bulk of mathematicians study mathematics because it is an inherently beautiful and logical enterprise. The rest of us use mathematics because it is a very powerful tool that helps us to evaluate situations and solve problems. Being able to make quantified statements allows us to predict phenomena with numerical precision and account for much of the behavior of the world. The goal of this unit is to re-familiarize you with basic algebra, introduce dimensional analysis, and teach you how to keep track of the quality of your measurements.

Algebra

One of the greatest tools of the scientist is an equation. A single equation can summarize millions of scientific observations or allow a scientist to accurately predict a new observation. It is very important to be able to perform simple algebraic operations and to use graphs.

Recall the basic equation of a line:

$$y = \mathrm{m}x + \mathrm{b}.$$

Variables y and x are related to each other. The constants m (slope) and b (y-intercept) describe the details of the relationship. Features of the relationship are:

If you exactly know two sets of x and y values, you can calculate m and b.
If you know m, b and the value of either variable (x or y), then you can predict the value of the other variable.

Simpler equations that are frequently useful are direct proportions and indirect proportions. They can be expressed in ratios or a single equation. If two variables are related by a direct proportion, then as the value of one variable increases, the value of the other variable also increases. If two variables are related by an indirect proportion (also called an inverse proportion), then as the value of one variable increases, the value of the other one will decrease.

	Direct Proportion	Indirect Proportion
Ratio form	$\dfrac{a_1}{a_2} = \dfrac{b_1}{b_2}$	$\dfrac{a_1}{a_2} = \dfrac{b_2}{b_1}$
Equation form k is a constant	$a = kb$	$a = k\dfrac{1}{b}$ or $ab = k$

Like the general equation of a line, the mathematical information in direct or indirect proportions can also be expressed in graphical form.

Many algebraic transformations and manipulations are performed by changing both sides of an equation by using identical procedures. If an equation is correct, then you can add or subtract a value from both sides and the result of your operation will still be correct. Likewise, multiplying or dividing both sides of an equation by the same number will not change its truthfulness, but may change its usefulness.

Dimensional Analysis

Dimensional analysis is a **very** powerful method for changing the form of a number without changing its value. If two quantities have different units, then comparing them might be difficult. For instance, it is not obvious whether 2300 g or 4 lb is a larger mass.

Dimensional analysis relies on equivalence statements. If you know that two quantities are equivalent, then you can build conversion factors from the equivalence statement by placing one side of the equality over the other to form a ratio. Conversion factors can transform a quantity without changing its value:

An equivalence statement: 3 dozen units = 36 units

Conversion factors: $\dfrac{3 \text{ dozen units}}{36 \text{ units}}$ or $\dfrac{1 \text{ dozen}}{12}$ or $\dfrac{12}{1 \text{ dozen}}$

The usefulness of conversion factors is that their numerical value is 1. The ratio of any two quantities that are equal is 1. That means that we can multiply any quantity we want by a conversion factor to change its form without changing its value.

Let's go back to the comparison of masses. When we use conversion factors, we need to do arithmetic analysis and unit analysis. If we transform the mass 2300 g to pounds, then we can compare the two masses much more easily. If you look in your textbook, you can find that 454 g = 1 lb. To change 2300 g to pounds, we will build the conversion factor that puts g in the denominator, so that we can divide out the undesirable unit and replace it with pounds for comparison to the quantity 4 lb:

$$2300 \text{ g} \times \frac{1 \text{ lb}}{454 \text{ g}} = 5.066 \text{ lb}$$

Since 5 lb is clearly greater than 4 lb, we know that 2300 g (5.066 lb) is the greater mass.

It is imperative that you always express quantities with both a number and a unit and that you always keep track of units during calculations. This is one of the best ways to prevent careless errors!

Density

A specific example of a conversion factor that is important to chemists is density. It facilitates conversions between the known volume of a material and its mass. Density is the ratio of the mass of a material to its volume and typically has units of g/mL. You can look up the density of common materials in your textbook or a reference work like the *CRC Handbook of Chemistry and Physics*. For example, the density of aluminum is 2.70 g/mL. Previously, we used an equivalence statement to construct a conversion factor. Likewise, we can use a conversion factor to construct an equivalence statement. Since the density of aluminum is 2.70 g/mL, we know that

$$2.70 \text{ g} = 1 \text{ mL}.$$

If we know that we have a 245 g block of aluminum, we could do the following calculation to determine the volume it occupies:

$$245 \text{ g} \times \frac{1 \text{ mL}}{2.70 \text{ g}} = 90.7 \text{ mL}.$$

Precision in Decision Making

The precision of a measurement is an indication of how reproducible it is. If you needed to decide whether a jockey or an NFL lineman were heavier, you would not need to know their weights very accurately. Jockeys weigh close to 100 pounds, but NFL linemen weigh approximately 300 pounds. For that comparison, you probably would not even need a balance. If you wanted to know which of your 1985-minted pennies were heavier, you would need to be very confident in your measurement. You might need to know the mass of each penny accurately to 5 digits to know which penny is heavier, as their masses are very close. To correctly make a difficult decision, you need reliable and reproducible data.

Significant Figures

The term "significant figure" describes any digit in a measured or calculated value that contains a value of which we are certain. If we reliably know many digits in a number, the number is quite precise and can be used to make very demanding decisions. In trying to determine which digits are significant, one problem arises: The number "zero" in normal notation can either be used to represent the number zero or it can be used to just hold a place. If someone wrote that they had 2.00000 grams, then we would know that they had 2 grams and very little more. Nobody would write all those zeroes unless they were sure that they were reliable. If you had a less precise scale that couldn't measure past the first decimal place, you would write 2.0 grams. A truck scale might only be able to measure in thousands of pounds. A truck that weighed 3000 pounds might be indistinguishable from a truck weighing 3154 pounds on that scale. If you wanted to write "about three thousand pounds" using normal notation, you would be forced to write "3000 pounds". You would have no way to convey the possibility of uncertainty in the zeroes that follow the 3. People would not be sure whether you meant 3000 pounds within a pound or two or whether you meant 3000 pounds within 200 pounds. To keep the 3 in the thousands place, you had to write three zeroes as placeholders. In cases such as these, we need to be conservative and assume that the trailing zeroes are not significant.

We can be certain of the digits in a reported answer if the digits contain

Any nonzero digit,
Zeroes between non-zero digits, or
Zeroes that follow the decimal place AND a non-zero digit.

Examples include
1.04   3 significant digits,
0.00**450**   3 significant digits, and
301300   4 significant digits.

There is a special class of numbers that we will refer to as "exact numbers" because they are known exactly and have no error associated with them. They can be reported to an infinite number of significant figures. Because of this, exact numbers will never limit the number of significant figures in an answer. Since no measurement is ever exact, exact numbers do not result from measurements. They are the result of counting or a definition. You can have exactly 400 pennies, which is exactly 4 dollars. Scientists have agreed that 1 inch is exactly 2.54 cm. Good reference manuals will designate which numbers are exact and which are not.

During mathematical operations, we will need to keep track of the digits of an answer if we are confident in that answer. There are two simple rules that we will use. For **multiplication and division**, the number of significant figures is important.   The number in the calculation with the **fewest number of significant digits** limits the number of significant digits in our reported answer. For **addition and subtraction**, we need to be aware of the **rightmost significant digit** in each number of the calculation. The rightmost significant digit in a reported answer is the one that results from the addition or subtraction of significant digits. If we add a digit in which we are certain to a digit in which we are not certain, we lose the certainty of the result. Thus, the resulting digit in the answer will not be significant.   During this week's workshop, you will have the opportunity to see these rules in action and discuss them. You might be tempted to use the simpler rule for multiplication and division in all cases, but this will lead to trouble, as you will see when you do the cooperative group problems in this unit.

Examples

Multiplication and division:

$$\textbf{7.0} \times \textbf{2.31} = \textbf{16}.17.$$

Since 7.0 has two significant figures and 2.31 has three significant figures, the answer will be limited to 2 significant figures (the lesser number of significant figures). Thus, we would report the calculated answer 16.17 as just 16 if this were our final answer.

Addition and subtraction:

$$
\begin{array}{r}
74.215 \\
- \ 73.52 \\
\hline
0.695
\end{array}
$$

In this case, we are concerned with the rightmost digit in which we are certain. The rightmost column in which we know both digits is in bold. This column and all of the digits to the left of this digit will be significant in our answer. Thus, despite having 5 significant digits in the first number and 4 significant digits in the second number, we only have 2 significant digits in our answer. We lost the tens place and the ones place in our answer during the subtraction.

Scientific Notation

The potential uncertainty of a number like 3000 caused mathematicians to develop a new way to report numbers. If someone told you they had 3000 dollars, you might not know whether they had "about 3000 dollars" or "exactly 3000 dollars" because the trailing zeroes might just be placeholders. Scientific notation involves the reporting of numbers as a product of a number between 1 and $9.999\overline{9}$, and 10 to an appropriate power. The numbers 3×10^3,

3.0×10^3, 3.00×10^3 and 3.0000000×10^3 all have the same numerical value, but are expressed with different numbers of significant figures. We know that the last of those values was measured to a high level of precision and we can use it for more difficult decision-making.

Examples:

Normal Notation	Scientific Notation
0.0034	3.4×10^{-3}
34000	3.4×10^4
1765716	1.765716×10^6

You should notice that when 0.0034 was transformed to a number with only one digit to the left of the decimal place, we made it 1000 times larger. To restore the original value of 0.0034, 3.4 needs to be multiplied by 1/1000, which is equal to 10^{-3}.

Pre-workshop Problems

1. Underline each significant figure in the following numbers. Count the number of significant figures in each number.

a. 3400 m

b. 0.0002105 cm

c. 0.005000 s

d. 3040.0 lbs

2. Write each of the following numbers in scientific notation.

a. 3400 m

b. 0.0002105 cm

c. 0.005000 s

d. 3040.0 lbs

3. Convert each quantity to the specified unit.

a. 2.5 m to inches

b. 31 ounces to grams

c. 1.46 m to μm. (You should use your textbook if you do not know the equivalency of the μ symbol, short for 'micro.')

Cooperative Group Problems

1. Which car has the bigger engine, the 1994 5.7 L Trans Am or the 1989 Formula 350? Hint: What does the "350" stand for?

2.   Report each answer to the proper number of significant figures. Be careful to use the correct rule for each operation.

a. $9.2 + 2.2 =$

b. $32 \times 21 =$

c. $(33.45 \times 2.961) - 100.6 =$

d. $(98.5 + 6.2) \times 1.0031 =$

3. For each quantity, how many significant figures should you use to report the answer? Your answer will depend on how you obtain the value you report to your group.

a. The number of fingers on your right hand

b. Your height in inches

c. Your current checking account balance

4. On the graph paper on the following pages, draw graphs that represent

a. A direct proportion

b. An indirect proportion

c. The equation $y = -2.3x + 7$. When x is 12.4, what is y?

5. For the following equation, what is the value of T_f?

$$4.5 \times 0.9 \times (T_f - 100) = 100 \times 4.184 \times (T_f - 23).$$

Practice Exam Questions
Unit 1

1. The number 0.00527 expressed in scientific notation is

    a. 5.27×10^3
    b. 0.527×10^{-3}
    c. 5.27×10^{-3}
    d. 5.27×10^{-4}
    e. 5.27×10^{-2}

2. The number 1.05×10^{-3} g can also be written as

    a. 1.05 Mg
    b. 1.05 kg
    c. 1.05 pg
    d. 1.05 mg
    e. 1.05 µg

3. The number 6.27×10^{-4} can also be written as

    a. 0.0627
    b. 0.00627
    c. 0.000627
    d. 0.0000627
    e. 62700

4. What is the result of the following multiplication problem expressed in scientific notation to the correct number of significant figures?

    $(5.34 \times 10^8)(2.3 \times 10^{-6}) =$

    a. 1.23×10^3
    b. 1.23×10^{-3}
    c. 1.2×10^3
    d. 1.2282×10^3
    e. 1×10^3

5. What is the result of the following summation expressed to the correct number of significant figures?
9.77 + 4.671 =

    a. 14
    b. 14.4
    c. 14.44
    d. 14.40
    e. 14.441

6. Convert 5.43 kg to pounds
(1 lb = 453.6 g).

    a. 12.0 lb
    b. 12.0×10^{-5} lb
    c. 2.46×10^{6} lb
    d. 0.0120 lb
    e. 2.46 lb

7. Convert 3.30 cm^3 to mL.

    a. 0.00330 mL
    b. 3.30 mL
    c. 0.330 mL
    d. 330.0 mL
    e. 3300.0 mL

8. Convert 4.50 cm^3 to L.

    a. 4.50×10^{-3} L
    b. 0.450 L
    c. 4.50 L
    d. 4.50×10^{3} L
    e. 0.450×10^{3} L

9. Convert 62°C to Kelvin.

    a. –211K
    b. 62K
    c. 54K
    d. 335K
    e. 0K

10. The mass of a piece of silver is   12.4 g. The density of silver is 10.5 g/mL. What is the volume of this piece of silver?

    a. 1.181 mL
    b. 0.847 mL
    c. 130 mL
    d. 1.30 mL
    e. 1.18 mL

11. Convert 50.00 kJ into calories.
4.184 J = 1 calorie.

    a. 11.95 cal
    b. 11950 cal
    c. 0.01195 cal
    d. 0.2092 cal
    e. 209200 cal

12.   The height of the lab assistant is 62.5 in. What is her height in centimeters?

    a. 24.6 cm
    b. 15.9 cm
    c. 159 cm
    d. 246 cm
    e. 1590 cm

13. What is 21.5°C in the Fahrenheit system?

    a. 70.7°F
    b. 6.7°F
    c. −18.9°F
    d. −5.83°F
    e. none of the above

14. What is 25.1 cm^2 in square inches?

    a. 9.88 in^2
    b. 3.89 in^2
    c. 63.8 in^2
    d. 162 in^2
    e. 76.1 in^2

Unit 2:
The Nature of Matter and Its Changes

Classification schemes can be very valuable tools for organizing information, which is a very important activity for scientists. Studying classification schemes and the language imbedded in them can reveal important facets of a subject. The way that chemists think is frequently exposed in the study of the language that chemists use. In studying the definitions that are the foundation of classification schemes, you should keep in mind that these definitions were arbitrarily made by a single individual or small group of individuals. Over time, the usefulness of certain definitions has caused them to be accepted worldwide.

States of Matter

You probably started studying the different states of matter a long time ago and retain much of what you have learned, which will help you greatly in your study of chemistry. Three of the states of matter are solids, liquids, and gases. The components (atoms, molecules, or ions) of solids have very little kinetic energy relative to the energy of attraction between them. This causes the attractions to dominate, pulling the particles together into closely packed, organized structures. It is important to realize that, in general, materials in the solid state are the only materials that are organized. Adding heat to a solid causes an increase in the energy and motion of the components until they can break free of their rigid organization. The attractions still hold the components close enough together to be touching in the liquid state, but now they can slide past each other. You may remember that liquids have fixed volumes, but can adopt the shape of their container. Adding more energy to a substance can cause components to completely overcome the attractive force between them. Then the components of a substance will separate such that there are great distances between them in the gaseous state. This causes them to occupy large volumes. It is beneficial if you can draw pictures representing each state of matter as well as provide accurate definitions for each state of matter. Less familiar states of matter are plasma and supercritical fluids, which you will learn about in more detail if you continue to study chemistry and physics.

Homogenous vs. Heterogeneous

The definition of homogeneous is deceptively simple. Homogeneous means "uniform throughout". The deception lies in the application of the definition.

Homogenized milk is certainly uniform throughout. The milk at the top looks, smells, and tastes exactly like the milk at the bottom. Still, chemists consider milk to be heterogeneous because someone discovered that the fat molecules in the milk are floating in little drops within the solution. If you could look closely enough, you would be able to see the lack of uniformity of the milk.

Any chemist would say that an ingot of silver is pure and homogeneous. Again, there is an arbitrary application of the definition. If you look "too closely," you would be able to see that each silver atom contains very dense nuclei with positive charges. Very light, negatively charged electrons surround these very dense particles. Any atom is heterogeneous if you examine it on a small enough scale, so you might argue that any matter is not homogeneous. Although correct, this is not a very useful idea for chemists.

Chemists think about matter on a certain scale. It will be beneficial to you to learn how to think about matter in the same way during your study of chemistry. Chemists primarily study how the molecules and atoms are organized, not how things look to the naked eye or how the particles smaller than atoms are organized. A chemist uses the word "uniformity" to mean that the molecules and atoms are uniformly distributed. As you attempt to determine whether a substance is uniform throughout or not, you must know what molecules or atoms compose it, as well as how they are organized.

Pure Substances vs. Mixtures

Classification according to purity is another area where the thinking of a chemist is very different than the thinking of a non-chemist. In the world of the supermarket, 100% pure maple syrup makes perfect sense. It refers to maple syrup to which nothing has been added. In the world of a chemist, maple syrup cannot be a pure substance because it is a mixture. It is a mixture of sugars, water, and the interesting molecules that give maple syrup its distinctive flavor and color. *A chemist thinks of a pure substance as something that can be described using a single, invariable chemical formula.* The definition of a pure substance is "a substance with constant composition." The single chemical formula describes the constant composition. By contrast, a mixture has variable composition because it contains at least two different substances that can be present in varying proportions. Some "pure maple syrups" have more or less water in them than others, thus they can taste stronger or less strong.

Until you know something about the chemical composition of different substances, deciding whether they are pure will be difficult and possibly very frustrating. The chemical community generally understands "sand" to be SiO_2 and considers sand to possibly be a pure substance, depending on its preparation. Limestone is another possibly pure substance, also known as calcium carbonate ($CaCO_3$). As you learn more about the composition of your world, you will find it easier to decide whether things are pure substances or mixtures, just as you will more easily be able to predict whether a chemist will consider them to be homogeneous or heterogeneous.

Chemical Processes

Just as understanding how chemical formulas describe substances can help us decide whether a material is pure or not, it can help us to classify whether a material undergoes a chemical process or not. If the chemical formula of a substance does not change during a process, the process is a physical process. If the formula of a material does change during a process, the process is a chemical process. In a chemical process, at least one of the bonds between atoms will change (be formed or broken). Physical processes, on the other hand, do not change the identity of a material; they change its physical appearance. Because you may not yet have learned much about the nature of matter, deciding which processes are chemical or physical might be difficult. As you learn more about materials and practice making these types of decisions, chemistry will rapidly become more comfortable for you to discuss.

Heat and Temperature Changes

If you add heat to a substance, it is likely to rise in temperature. The amount of the temperature change depends on the mass and type of material being heated. Each material has a specific heat capacity (also called specific heat). This is the amount of heat required to raise one gram of the material by one degree Celsius. You can think of the specific heat capacity as the conversion factor that relates the heat change, the amount of material, and the temperature change. The units for specific heat are J/(g °C) or cal/(g °C). The equation used to quantify heat and temperature changes is

$$q = mC_{SH}\left(T_f - T_i\right),$$

where q is the heat change, m is the mass of material, C_{SH} is the specific heat, and T_f-T_i is the change in temperature, also expressed as ΔT.

Pre-workshop Problems

1. Using your textbook, write down definitions for each of the following terms:

    a. Pure substance

    b. Compound

    c. Mixture

    d. Element

    e. Atom

    f. Molecule

    g. Solution

    h. Isotope

    i. Alloy

    j. Homogeneous

    k. Heterogeneous

    l. Physical process

    m. Physical property

    n. Chemical process

    o. Chemical property

2.   Fill in the following table using words that you have defined in
Question 1. Do not use your textbook.

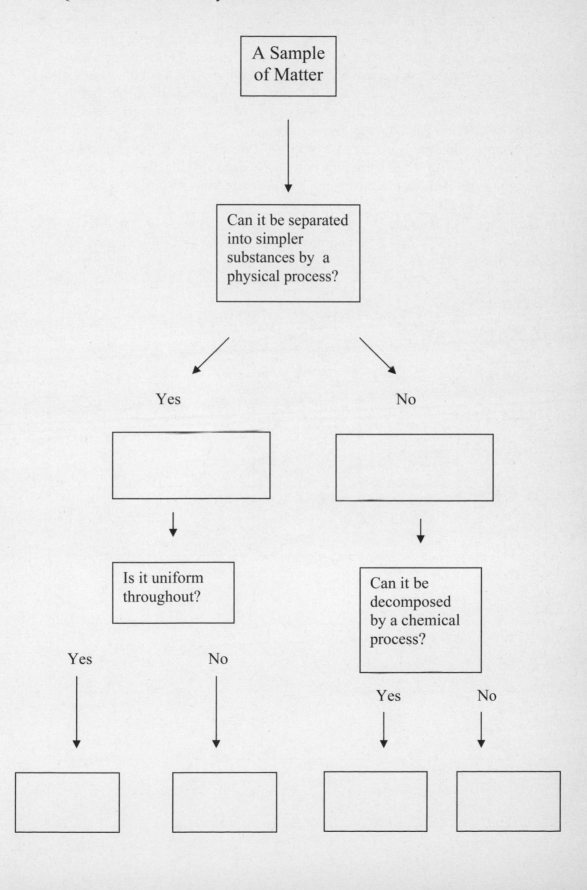

Cooperative Group Problems

1. Split your group into two subgroups. One of the subgroups will be the black subgroup; the other will be the red subgroup. Your workshop leader will give you a set of plastic bags containing assorted paperclips attached in various ways. These are a chemist's representation of various classes of matter. Decide how to describe the contents of each bag. First, list the number and type of each paperclip. Also calculate the percent composition of the bag for each constituent. Then decide which term (element, compound, mixture, or pure substance) best fit the contents of each bag. After your subgroup has finished describing the contents of each bag, re-form your entire group and compare your results.

Bag color (red or black)	Contents	Percent composition	Descriptive Terms	Possible Chemical Formulas
A				
B				
C				
D				
E				
F				
G				
H				

2. For this question, work in pairs. Discuss the following substances, and classify them using the vocabulary we have discussed in this unit. When possible, write chemical formulas for each substance. If you are unfamiliar with the materials listed, consult your text book. Can you think of an easy way to separate the mixtures?

   a. Air

   b. Water

   c. Aqueous hydrochloric acid

   d. Milk

   e. Brass

   f. Lime (the building material, not the fruit)

   g. Lemonade

   h. Sugar

   i. Table salt

   j. Nitrogen

   k. Rusted iron

   l. A mixture of sugar and broken glass

3.   Draw three pictures to represent sulfur dioxide. Draw one picture to
        represent the solid state, one to represent the liquid state, and one to
        represent the gaseous state. List the features that are illustrated in each
        drawing.

4. Think of at least eight physical processes that are familiar to you. Explain why they are physical processes. Now think of eight processes that are chemical processes, and explain why they are chemical processes. One particularly difficult example is the hardening of cement. In which class does it belong?

5. A 0.500 g sample of mercury (specific heat capacity = 0.14 J/g °C) cools from 27.0°C to 10.0°C. What heat change did the mercury undergo? Does the sign of the heat change make sense to you?

Practice Exam Questions
Unit 2

1. Which of the following is not a physical change?

   a. iron melting
   b. steel wool forming rust
   c. ice melting
   d. water boiling
   e. naphthalene evaporating

2. A liquid

   a. has a definite shape but not a definite volume
   b. does not have a definite shape and does not have a definite volume
   c. does not have a definite shape but has a definite volume
   d. has a definite shape and a definite volume
   e. none of the above

3. The state of matter for an object that has no definite volume and no definite shape is

   a. solid
   b. gas
   c. liquid
   d. mixed
   e. elemental

4. Wood burning on a campfire is an example of

   a. a physical change
   b. a precipitation reaction
   c. a compound
   d. a chemical change
   e. a pure substance

5. A sample of aluminum (specific heat capacity = 0.89 J/g °C) is heated from 22°C to 75°C using 54.0 J of energy. What is the mass of this sample of aluminum?

   a. 0.11 g
   b. 1.1 g
   c. 0.91 g
   d. 9.1 g
   e. 11 g

6. If a 5.4 g piece of iron is heated from 25°C to 75°C using 128.8 J of energy. What is the specific heat capacity of iron?

   a. 0.45 J/g °C
   b. 4.5 J/g °C
   c. 2.2 J/g °C
   d. 0.22 J/g °C
   e. 13 J/g °C

7. A 6.50 g sample of silver (specific heat capacity = 0.24 J/g °C) is heated with 25.6 J of energy. If the initial temperature of the silver was 22.0 °C what is the final temperature?

   a. 6.0 °C
   b. 23.0 °C
   c. 38.4 °C
   d. 671.3 °C
   e. 715.3 °C

8. Sand is an example of

   a. a compound
   b. a solution
   c. a mixture
   d. a pure substance
   e. an element

9. An example of a substance or type of substance that might not always have the same composition is

    a. propane
    b. a compound
    c. a pure substance
    d. an element
    e. a solution

10. Water from Lake Arrowhead is an example of

    a. a mixture
    b. a pure substance
    c. a compound
    d. an element
    e. none of the above

Unit 3: Building Blocks of the Universe

My high school chemistry teacher used to refer to the periodic table as "God's tinker toy set." He claimed that if we knew everything about the elements in the periodic table, then we could build anything we wanted. Although exaggerated, there is some simple truth to his claims. Everything in your life is a chemical and is composed of atoms in various arrangements. If you wish to be able to understand and control your world, you should understand the elements and their fundamental properties.

The Periodic Table of the Elements

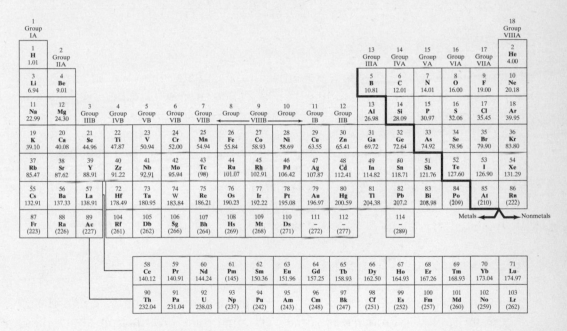

Figure 3.1 Periodic Table of the Elements.

The development of the periodic table took chemists and physicists a very long time to organize and to understand how to use it. During the rest of this course, you will use this tool to predict which substances are stable, to understand similarities and differences between elements, and to convert between the mass of a substance and the number of particles it contains.

Groups

In the periodic table, a group is the set of all the elements in the same column. These elements have similar reactivity and other similar properties.   Learn the group names.

Group name	Group number	Compound with hydrogen	Typical charge of ions	Example
Alkali metals	1	NaH	+1	Na^+
Alkaline earth metals	2	MgH_2	+2	Mg^{2+}
Pnictogens	5 (15)	NH_3	-3	P^{3-}
Chalcogens	6 (16)	OH_2	-2	S^{2-}
Halogens	7 (17)	FH	-1	Cl^-
Noble gases	8 (18)	None	None (0)	Ne

Atoms

Dalton was one of the first scientists to postulate that any element was composed of individual building blocks called atoms. He believed that an atom was indivisible and that all atoms of a single element were identical. According to Dalton's Atomic Theory, these atoms could combine in different ratios to form compounds, which have different properties than either element does individually. Although he was right about the formation of compounds, other scientists would later learn that 1) atoms are composed of smaller particles and can be decomposed, and 2) two atoms of the same element can be different, if they have different masses.

The Subatomic Particles

The three basic building blocks of the atom are the proton, neutron and electron. Table 1 summarizes some very important information about these particles.

Table 1

Name	Symbol	Charge	Relative mass	Mass
proton	p	+1	1	1.67×10^{-27} kg
neutron	n	0	1	1.67×10^{-27} kg
electron	e	-1	0.0005 (1/2000)	9.1×10^{-31} kg

The Identity of an Atom

An atom's fundamental behavior is determined by the amount of positive charge in its nucleus. Thus, if we know how many protons are in the nucleus of an atom, we can determine its identity. The number of protons in an atom or ion is called its *atomic number* and is given the symbol Z.

Charge Balance

During *normal* chemical reactions, the nucleus of an atom does remain unchanged and indivisible. Electrons, on the other hand, can be gained or lost to form ions, which can be positively or negatively charged. An atom is neutral because it has an equal number of protons and electrons. An atom that loses an electron will have a net +1 charge because it has lost a negative charge, while an atom that gains an electron gains a negative charge and thus will have a net −1 charge. Positively charged ions are called cations. Negatively charged ions are called anions. A mathematical way to determine the charge of an atom or ion is to subtract the number of electrons from the number of protons.

Mass Number of Atoms

Upon examination of Table 1, you can see that the overall mass of an atom depends primarily on the number of protons and neutrons it contains; the electrons are so light that they can be ignored for most practical purposes. The mass number of an atom is equal to the total number of protons plus the neutrons it contains; it is given the symbol A.

NOTE: The periodic table has information about the average mass of an element, but does not have information about the mass number of a particular atom or isotope.

Elemental Symbols

A complete symbol contains information about the elemental symbol (X), atomic number (Z), mass number (A), and charge (n). The general format of the symbol is shown in the following diagram:

$$_Z^A X^n$$

Isotopes

If two atoms have the same atomic number but different mass numbers, then they are clearly not identical. Even so, they will still have the same basic chemical reactivity and will form compounds with the same formulas. Carbon-12 is an atom that has 6 protons and 6 neutrons, which results in a total mass of 12. Carbon-13 also has 6 protons (which is why it is a carbon atom), but it has 7 neutrons. Both carbon-12 and carbon-13 will form carbon monoxide or carbon dioxide on reaction with oxygen. One of the most interesting applications of chemistry is the dating of artifacts by determining the amount of radioactive carbon-14 that they contain.

Predicting Charges

For reasons that you will learn later, any atom or ion having the same number of electrons as a noble gas will be stable. Metal atoms, which are on the left side of the periodic table, can easily lose a few electrons to become stable ions. Sodium atoms tend to lose one electron, and magnesium atoms tend to lose two electrons so they will have ten electrons just like neon. Oxygen will gain two electrons and fluorine will gain one electron to become stable ions. Take a look at the periodic table and re-read this section to make sure you understand how to predict charges of simple ions.

Charge Balance in Ionic Compounds

Although they are called "stable ions," neither F^- (fluoride) nor O^{2-} (oxide) can be isolated in bottles as pure substances. To be stable enough to be isolated in large quantities, substances must have a zero net charge. For this reason, cations and ions will combine in fixed ratios so that their charges cancel. A magnesium ion (Mg^{2+}) will require two fluoride ions (F^-) to form the stable MgF_2 compound. The formula MgF_2 means that one magnesium ion is combined with two fluoride ions.

Polyatomic ions

In addition to simple ions that are formed from an atom that either gains or loses electrons, there are many polyatomic ions. In the next unit, we will learn both the naming system for these ions and .their formulas; for now, you can simply look them up in your book. They also need counterparts of opposite charges to form stable species that can be put into a bottle.

Pre-workshop Problems

1.   Write down the names of the halogens.

2.   You will need to know the names and symbols for all of the elements if you are going to be able to understand what a chemist says and do well in this course. It helps to group them by their first letters before studying them.

For each letter below, make a list of the names of all symbols that start with that letter and their names. Study all of the names and symbols before your next team meeting.

a. S

b. C

c. A

3.   How many protons, neutrons, and electrons are contained in a +2 ion formed from a platinum-196 atom? Write mathematical equations that prove you are correct.

Cooperative Group Problems

1. Each member of your team should spend 5 to 10 minutes drawing representations of boron-10 and boron-11 atoms. Then re-form the group and decide whose drawings are the best and why. Compose a "best of the best" drawing of the two isotopes.

2. Complete the following table. Each member should fill in one box. Take turns until the table is complete.

Element	Sb			Kr
Atomic number		14		
Mass number	121			
Nuclear charge			6	
# of neutrons		14	7	56
# of electrons			6	36
Charge	-3	+4		

3. Your workshop leader will provide each of you with a small stack of elemental symbol flash cards and explain the game "around the world". Play this game to learn the elemental symbols and names. If you study before your group meets, you will be much more successful.

4. To complete the following table, write the formula of the compound formed by the combination of the cation in the top row with the anion in the left column. The first box is completed as an example. Have each person complete one box until all boxes are filled.

	Na^+	Hg_2^{2+}	Fe^{3+}	Ba^{2+}	Cu^+
Cl^-	NaCl				
NO_3^-					
SO_4^{2-}					
N^{3-}					
CO_3^{2-}					

5. Names of common cations and anions are given in the top row and left column of the following table. You may use your textbook to determine the formulas of the complex anions. Fill in the chemical formulas of the stable compounds formed from each combination of ions in the table.

	Fluoride	Phosphide	Selenide	Chromate	Phosphate
Calcium					
Sodium					
Iron (III)					
Lead (IV)					
Aluminum					

Practice Exam Questions
Unit 3

1. The symbol Co stands for what element?

    a. copper
    b. chlorine
    c. carbon
    d. cobalt
    e. californium

2. The symbol for the element silver is

    a. Si
    b. Ag
    c. S
    d. As
    e. Sc

3. How many protons, electrons, and neutrons, respectively, does ^{15}N have?

    a. 7, 7, 8
    b. 7, 7, 7
    c. 7, 14, 7
    d. 7, 8, 6
    e. 7, 6, 8

4. Predict the formula of the compound that is composed of the ions Mg^{2+} and Cl^-.

    a. Mg_2Cl
    b. $MgCl$
    c. Mg_2Cl_2
    d. $MgCl_2$
    e. Mg_4Cl_2

5. The chemical formula, $Al(NO_3)_3$, indicates

a. one atom of aluminum, one atom of nitrogen, and 3 atoms of oxygen
b. one atom of aluminum, one atom of nitrogen, and 6 atoms of oxygen
c. one atom of aluminum, one atom of nitrogen, and 9 atoms of oxygen
d. one atom of aluminum, three atoms of nitrogen, and 6 atoms of oxygen
e. one atom of aluminum, three atoms of nitrogen, and 9 atoms of oxygen

6. The elements I and At fall under what group name?

    a. Alkali metals
    b. Alkaline earth metals
    c. Transition metals
    d. Halogens
    e. Noble gases

7. The number of protons in an element is given by

    a. mass number
    b. atomic number
    c. group number
    d. period number
    e. isotope number

8. The difference between the mass number and the atomic number gives the

    a. number of electrons
    b. number of protons
    c. number of neutrons
    d. period number
    e. group number

9. The symbol Sn represents the element

    a. Samarium
    b. Antimony
    c. Silicon
    d. Tin
    e. Scandium

10. Predict the formula of the compound that is composed of the ions Fe^{3+} and O^{2-}.

    a. FeO
    b. FeO_2
    c. Fe_2O
    d. Fe_2O_3
    e. Fe_3O_2

11. Predict the formula of the compound that is composed of the calcium and iodide ions.

    a. CaI
    b. Ca_2I
    c. Ca_2I_3
    d. CaI_2
    e. Ca_3I_2

12. Predict the formula of the compound that is composed of the calcium and nitride ions.

    a. CaN
    b. Ca_2N
    c. Ca_2N_3
    d. CaN_3
    e. Ca_3N_2

13. How many protons, neutrons, and electrons, respectively, compose the ion formed from the calcium-38 atom?

    a. 20, 20, 20
    b. 20, 18, 20
    c. 20, 18, 18
    d. 20, 20, 22
    e. 20, 18, 22

14. The element most similar to Si is

    a. S
    b. Al
    c. As
    d. Ge
    e. Ar

Unit 4: What's in a Name?

I have heard people say that the average Biology or Chemistry class has as much new vocabulary as a typical French class. Obviously, this will mean that you should treat science classes in much the same way that you treat language classes. There are numerous words that you must learn. After you have learned the meanings of the words, then you can start to use them in meaningful ways. Until you learn how to name chemical compounds, you will find it impossible to understand a chemist or your instructor.

Polyatomic Ions

Your textbook certainly contains a table of polyatomic ions. A polyatomic ion is a cluster of atoms that are bound to each other into a stable unit that carries a charge. As chemists discovered polyatomic ions, they tried to invent names that make sense. You should start memorizing the names and formulas of the common polyatomic ions as soon as possible.

Common Oxyanions

As people isolated ores from the ground and analyzed the composition of the world, they noticed that many substances were composed of a metal, a nonmetal, and a certain number of oxygen atoms. Each nonmetal regularly was found in combination with a certain, specific number of oxygen atoms. The ionic non-metal/oxygen combination became known as an oxyanion and has a certain negative charge associated with it. The most common oxyanions were given a name with a prefix related to the non-metal and a suffix "ate." Unfortunately, there is no simple system to predict the number of oxygen atoms or the charge of an oxyanion. **Thus, the formulas of the "ates" must simply be memorized**.

Name	Formula
Nitrate	NO_3^-
Chlorate	ClO_3^-
Bromate	BrO_3^-
Sulfate	SO_4^{2-}
Phosphate	PO_4^{3-}
Chromate	CrO_4^{2-}
Dichromate	$Cr_2O_7^{2-}$
Carbonate	CO_3^{2-}
Manganate	MnO_3^-

Other Oxyanions

As chemists started to mess with nature, they soon were making lots of things that you would not be able to easily find in the natural world. They soon found or synthesized oxyanions with different numbers of oxygen atoms than the common oxyanions had. This meant they needed a way to name these new oxyanions. Luckily for us, any series of oxyanions of an element varies only in the number of oxygen atoms, not in the charge.

A representative series of anions illustrates the naming system. Notice that the simplest anion, one without any oxygen at all, takes an "ide" suffix.

Name	Formula	Prefix/Suffix	Number of oxygen atoms
Chloride	Cl^-	"ide"	None
Hypochlorite	ClO^-	"hypo" /"ite"	One less than "ite"
Chlorite	ClO_2^-	None /"ite"	One less than "ate"
Chlorate	ClO_3^-	None/"ate"	The common number
Perchlorate	ClO_4^-	"per" /"ate"	One more than "ate"

Hydrogen-containing Ions

For oxyanions with higher charges, stable ions can be formed from a combination of a hydrogen ion and the anion. Carbonate, CO_3^{2-} has a –2 charge. Hydrogen carbonate is HCO_3^-. The older name for this ion is bicarbonate. You should become familiar with the following table, which includes some of the hydrogen-containing anions.

Name	Formula
Hydrogen carbonate	HCO_3^-
Hydrogen phosphate	HPO_4^{2-}
Dihydrogen phosphate	$H_2PO_4^-$

Miscellaneous Ions

Your instructor may require you to know other important ions that do not follow the systematic nomenclature. Your textbook has a table of common ions that you should memorize. Cyanide (CN^-) is an ion that most instructors will expect you to know because it is famous as a poison and is extremely good at binding metal ions. Hydroxide (OH^-) is a very common anion and base.

Types of Compounds

Different types of compounds are named in different ways. Thus, you must become familiar with the types of compounds and the ways to name them. There are particular naming systems for salts of metals that can only take on one charge, salts of metals that can have different charges, acids, and molecular compounds.

Salts of Metals That only have One Stable Charge, Type I Compounds

The metals in Groups I, Group II, scandium, aluminum and zinc only have one possible stable charge. Group I metals always take on a +1 charge. Group II metals and zinc always take on a +2 charge. Aluminum and scandium always take on a +3 charge. In salts of these metals, there is only one possible formula of a stable compound with no net charge. For example, zinc chloride must be $ZnCl_2$ because zinc ions always have a charge of +2 and chloride ions always have a -1 charge. For salts involving these metals, naming the compound involves only naming the metal and the anion. $CaCO_3$ is named calcium carbonate. These compounds are frequently called Type I salts and the metals they contain are frequently called Type I metals.

Salts of Metals With Multiple Possible Charges, Type II Compounds

Lead can form either $PbCl_2$ or $PbCl_4$. Hence, the name lead chloride could refer to two different compounds. Because of this problem, Alfred Stock developed a naming system for these salts. With this system, we first list the cation name, then a Roman numeral identifying the charge on the metal, then the anion name. This system of naming is used for any metal salt that does not contain a Type I metal. The aforementioned lead compounds are named lead (II) chloride, $PbCl_2$, and lead (IV) chloride, $PbCl_4$.

Acids

Acids are named after the anion that is a component of the acid. Thus, if you want to name acids, you need to know the names of the anions. The most common acids are compounds composed of H^+ ions and stable anions. The following table of acids helps to illustrate the system; it uses the chlorine-containing anions that were listed earlier in this unit.

Name	Formula	Acid Formula	Acid Name
Chloride	Cl^-	HCl	Hydrochloric Acid
Hypochlorite	ClO^-	HClO	Hypochlorous Acid
Chlorite	ClO_2^-	$HClO_2$	Chlorous Acid
Chlorate	ClO_3^-	$HClO_3$	Chloric Acid
Perchlorate	ClO_4^-	$HClO_4$	Perchloric Acid

Molecular Compounds

Generally, neutral compounds involving nonmetals form molecular compounds. Naming and writing formulas for these compounds is easy because the naming system tells us exactly how many atoms are in the compound. At this point, you can imagine that a molecular compound is composed of exclusively non-metal elements. For example, N_2O_3 is dinitrogen trioxide. If a compound has both a metal and a non-metal, it will be a salt, not a molecular compound.

Pre-workshop Problems

1.   Find the list of common ions in your textbook. Write down the names of any cationic polyatomic ions.

2.   Write the formulas for each ion. Briefly explain how you figured out each formula.

a. sulfite

b. periodate

c. hypobromite

3.   For naming purposes, determine the type of each of the following compounds. Write the formula of each compound.

a. Sodium sulfate

b. Iron (III) chloride

c. Hypochlorous acid

d. Sulfur trioxide

Cooperative Group Problems

1. Look at the table of common polyatomic ions in your book.

Using the following example (for chlorine-based anions), create tables for anions containing N, P, Br, Cr, S, and C. Keep ions derived from the same element in the same table. Note that not every ion type exists for every element; for example, there is no "hypocarbonite" ion.

Chlorine-containing anions

Name	Formula	Prefix/Suffix	Number of oxygen atoms
Chloride	Cl^-	"ide"	None
Hypochlorite	ClO^-	"hypo" /"ite"	One less than "ite"
Chlorite	ClO_2^-	None /"ite"	One less than "ate"
Chlorate	ClO_3^-	None/"ate"	The common number
Perchlorate	ClO_4^-	"per" /"ate"	One more than "ate"

2.   One definition of an acid is "an acid is something that can donate a H^+ ion to an anion or neutral compound."   For each oxyanion that you place in your table for Question 1, there is a corresponding acid. To obtain the formula for these acids, you must combine the anion with enough H^+ to get a neutral compound. For example,

$$PO_4^{3-} + 3\ H^+ \rightarrow H_3PO_4.$$

Take turns filling in rows in the following table.

Anion	Acid Formula	Acid Name
PO_4^{3-}	H_3PO_4	Phosphoric acid

3. Your group leader has flash cards for nomenclature. Using these flash cards, play the game "Around the World" with your group. If your group cannot name the compounds or provide formulas for the names, take a break and review the types of compounds and naming systems before continuing. If you have been studying nomenclature this week, you should be a contender for victory.

Practice Exam Questions
Unit 4

1. A compound that is composed of two different elements is

    a. a binary compound
    b. a solution
    c. a mixture
    d. a molecule
    e. an atom

2. An ionic compound typically contains

    a. a metal and a metal
    b. a nonmetal and a nonmetal
    c. a transition metal and a transition metal
    d. a metal and a nonmetal
    e. covalent bonds

3. A compound that contains two nonmetals is a

    a. Type I binary ionic compound
    b. Type II binary ionic   compound
    c. molecular compound
    d. solution
    e. mixture

4. Which of the following is not a polyatomic ion?

    a. NH_4^+
    b. PO_4^{3-}
    c. OH^-
    d. H^+
    e. O_2^{2-}

5. _____ contain several atoms bound together with a unit charge.

    a. Compounds
    b. Allotropes
    c. Charged elements
    d. Polyatomic ions
    e. Acids

6. PCl_5 has the name

    a. phosphorus chloride
    b. phosphorus pentachlorite
    c. phosphorus chlorite
    d. phosphorus pentachloride
    e. phosphorus (V) chloride

7. The name dinitrogen trioxide has the formula

    a. $2 NO_3$
    b. N_2O_3
    c. NO
    d. N_3O_2
    e. N_6O_6

8. $Fe(NO_3)_3$ has the name

    a. Iron nitrate
    b. Iron trinitrate
    c. Iron (II) nitrate
    d. Iron (III) nitrate
    e. none of the above

9. Which of the following is named correctly?

    a. $Mn(OH)_2$, Manganese hydroxide
    b. Na_2SO_3, Sodium sulfate
    c. SF_6, Sulfur hexafluoride
    d. N_2O_5, Nitrogen oxide
    e. all of the above

10. The name for H_2SO_4 is

    a. Sulfurous acid
    b. Hydrogen sulfate
    c. Sulfuric acid
    d. Hydrogen sulfite
    e. Dihydrogen sulfur tetroxide

11. Chlorous acid is

    a. HClO
    b. H_2ClO
    c. $HClO_2$
    d. H_2ClO_2
    e. $HClO_3$

12. Which is not named correctly?

    a. $Fe(OH)_2$, iron hydroxide
    b. HBrO, hypobromous acid
    c. P_4O_{10}, tetraphosphorous decoxide
    d. PbO, lead (II) oxide
    e. CO_3^{2-}, carbonate

13. The formula for sodium phosphate is

    a. $NaPO_3$
    b. $NaPO_4$
    c. Na_2PO_3
    d. Na_3PO_3
    e. Na_3PO_4

14. The formula for ammonium carbonate is

    a. NH_3CO_3
    b. NH_4CO_3
    c. $(NH_3)_2CO_3$
    d. $(NH_4)_2CO_3$
    e. $(NH_3)_3CO_3$

15. $Ca(ClO_3)_2$ has the name

    a. calcium (II) chlorite
    b. calcium chlorite
    c. calcium (II) chlorate
    d. calcium chlorate
    e. calcium dichlorite

Unit 5: Describing the Action

As you will see during your study of chemistry, chemists try to describe the world, try to predict how components of the world interact with each other, and try to use this knowledge to improve the quality of life. Chemical reactions are involved in the manufacture of materials, the growth and death of all organisms, and many of the processes that generate energy. Understanding these reactions can help us to take control of them.

Conservation of Matter

During ordinary chemical reactions, we can observe two important events: (1) matter is conserved, and (2) atoms do not change identity. Because of these observations, we know that the number of each type of atom at the beginning of a reaction must be equal to the number of each type of atom at the end of a chemical reaction. Whenever we try to describe a process using a chemical equation, we need to keep this rule in mind.

Formulas

A formula tells us the composition of a material. First, it tells us which atoms are parts of the material; then, it tells us how many atoms of each type are included in the material. A very simple formula is H_2O, which is the formula for water. Water is composed of two hydrogen atoms attached to one oxygen atom. You could visualize this formula as two little spheres attached to one big sphere.

Figure 5.1 Water molecule.

Our visual representation of water will improve in accuracy as we learn more about the details of the shape of the molecule and its behavior.

Sometimes, the formula will tell us more about the arrangement of atoms in the material. Consider the formula $Ca(NO_3)_2$, which tells us that there are two NO_3^- units for each Ca^{2+}. The ions can be separated from each other fairly easily, but the NO_3^- ions will tend to stay in one piece.

Figure 5.2 shows a simple visualization of the components of calcium nitrate.

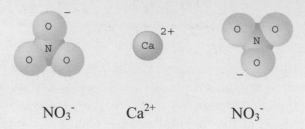

$$NO_3^-   \qquad   Ca^{2+}   \qquad   NO_3^-$$

Figure 5.2 Calcium nitrate.

Unbalanced Chemical Equations

Chemical equations describe chemical reactions. For example, if you mix sodium with oxygen, you will produce sodium oxide. This can be expressed concisely with the following chemical equation:

$$Na + O_2 \rightarrow Na_2O$$

This equation describes the identities of the reactants and products. However, as written, it suggests a violation of the Law of Conservation of Matter. Note that there was only one sodium atom in the reactant, but there are two sodium atoms in the product. We might be tempted to write NaO as a product to fix this problem, but we are not allowed to change any of the subscripts in the formulas. NaO does not exist, but Na_2O does. Remember that charges must balance to have a stable compound, so sodium oxide requires two sodium(+1) ions for every oxide(-2) ion.

Balancing Chemical Equations

In order to correctly describe what happens during a chemical reaction, we must correctly describe the identities of the reactants and products. Our chemical equation must also reflect the fact that the Law of Conservation of Matter is always obeyed. Thus, we must always have the same number of each type of element in the reactants as we have in the products. In order to do this, we will include coefficients as part of the equation. They will identify the relative amounts of reactants and products that are involved in the reaction.

Start by looking at one of the atoms in the equation and trying to add or change the coefficients of the compounds that contain it; your goal is to have the same number of that atom on each side of the equation. It is often helpful to look at the most complicated molecule in the equation first. If we change the coefficient of the Na_2O, we alter the number of sodium atoms and the number of oxygen atoms. To balance the number of sodium atoms, we must add a coefficient of 2 to the Na; then we will have 2 sodium atoms on each side of the equation:

$$\underline{2}\,Na + O_2 \rightarrow Na_2O$$

After we make this change, we must preserve the 2-to-1 ratio of sodium to sodium oxide; this preserves the equality in the number of sodium atoms on each side of the reaction.

Next, we need to consider the number of oxygen atoms. In the preceding reaction, there are two oxygen atoms on the left side of the equation, but only one on the right side. Thus, we will want to change the coefficient of the sodium oxide from one to two. When we do this, we must remember that changing the coefficient of the sodium oxide atom will alter the number of sodium atoms as well as the number of oxygen atoms. If we double the coefficient of sodium oxide, we must also double the coefficient of the sodium, as shown by

$$\underline{4}\,Na + O_2 \rightarrow \underline{2}\,Na_2O$$

Balancing a reaction is like a game: we must constantly be aware of what we have done in the past and how our actions will affect what we have done. As we learn chemistry, we will learn more strategies to help us balance equations.

Properly Balanced

Most chemists prefer balanced chemical equations to contain the lowest whole number coefficients that describe the proper ratios of reactants and products. A typical type of reaction is a combustion reaction. In a combustion reaction, a flammable substance is burned in the presence of oxygen to form oxides. If the flammable substance contains carbon, one of the products is carbon dioxide. If the flammable substance contains hydrogen, one of the products is water.

We will now show the stepwise balance of the equation for the combustion of ethanal, CH_3CHO. The first step is to write the unbalanced equation. Ethanal plus oxygen will form carbon dioxide and water:

$$CH_3CHO + O_2 \rightarrow CO_2 + H_2O$$

Again, we start with the complicated molecule; in this case, it is CH_3COH. All of the carbon in the starting material ends up in the carbon dioxide. Since there are two carbon atoms in the ethanal, then we need a coefficient of 2 for the carbon dioxide to balance the carbon atoms:

$$\underline{1} \, CH_3CHO \; + \; O_2 \; \rightarrow \; \underline{2} \, CO_2 \; + \; H_2O$$

Likewise, all of the hydrogen in the ethanal becomes part of the water. Since the ethanal contains 4 hydrogen atoms, we need a coefficient of 2 for the water:

$$\underline{1} \, CH_3CHO \; + \; O_2 \; \rightarrow \; 2 \, CO_2 \; + \; \underline{2} \, H_2O$$

After we have added that coefficient, we have balanced the number of carbon and hydrogen atoms, so only the oxygen balance is left. We have fixed all of the coefficients except that of the elemental oxygen, so we will add a coefficient to complete the balancing of the equation. Due to the carbon dioxide, there are 4 product oxygen atoms; due to the water, there are 2 product oxygen atoms. This results in a total of 6 product oxygen atoms. The ethanal supplies one oxygen atom, so the oxygen starting material must supply 5 oxygen atoms. Since one-half of an O_2 molecule equals one oxygen atom, 5/2 O_2 molecules equals 5 oxygen atoms. Thus, we can balance the equation by writing

$$CH_3CHO \; + \; \underline{5/2} \, O_2 \; \rightarrow \; 2 \, CO_2 \; + \; 2 \, H_2O$$

Chemists prefer to have integer coefficients for all reactants and products, so we will convert 5/2 to a whole number. If we multiply all of the coefficients by a factor of 2, we eliminate the fraction from our equation:

$$2 \, CH_3COH \; + \; 5 \, O_2 \; \rightarrow \; 4 \, CO_2 \; + \; 4 \, H_2O$$

Designation of States

The most important information that can be conveyed in a complete chemical equation is the state of each reactant and product. This is done using an abbreviation in parentheses. Solids are followed by an (s), liquids by an (l), gases by a (g), and solutes dissolved in water by an (aq) for aqueous. For completeness, we can re-write the balanced equations as

$$2 \, Na(s) \; + \; O_2(g) \; \rightarrow \; Na_2O(s)$$

$$2 \, CH_3COH(aq) \; + \; 5 \, O_2(g) \; \rightarrow \; 4 \, CO_2(g) \; + \; 4 \, H_2O(l)$$

Complex Ions

Polyatomic ions frequently do not get involved in a reaction. We can count them as units instead of keeping track of each individual atom that composed them. Thinking of groups of atoms can greatly simplify the balancing of some chemical equations. Consider the following unbalanced equation:

$$NaOH(aq) + H_3PO_4(aq) \rightarrow H_2O(l) + Na_3PO_4(aq)$$

Any chemist would recognize that phosphate (PO_4^{3-}) is unchanged during this reaction. This observation allows the counting of phosphate as a unit, which is much simpler than accounting for each individual atom. Initial inspection of the equation reveals that the phosphate ions are already balanced. Placing a coefficient of three in front of the sodium hydroxide will balance the sodium atoms:

$$\underline{3}\,NaOH(aq) + H_3PO_4(aq) \rightarrow H_2O(l) + \underline{1}\,Na_3PO_4(aq)$$

After considering the sodium atoms and phosphate ions, we only need to worry about the hydrogen and non-phosphate oxygen atoms. On the left, there are 6 hydrogen atoms; thus, a coefficient of three for the water molecules will balance the hydrogen atoms. This also balances the oxygen atoms:

$$\underline{3}\,NaOH(aq) + \underline{1}\,H_3PO_4(aq) \rightarrow \underline{3}\,H_2O(l) + Na_3PO_4(aq)$$

By evaluating numerous chemical equations, chemists have noticed many patterns of behavior that make balancing chemical equations easier. Treating polyatomic ions as groups is one such example. In the next unit, we will see that chemists also look for patterns of chemical reactivity. If you wish to do well in chemistry classes, discovering these patterns and thinking like a chemist will be a great asset to you.

Patterns of Chemical Reactivity

In the last example, we saw that it was helpful to think of the phosphate ion as a unit. If we examine the rest of the equation, we can see an important reactivity pattern. Hydrogen ions in the phosphoric acid react with hydroxide ions in the sodium hydroxide to form water. There are many examples where hydrogen atoms from one compound react with hydroxide atoms from another to form water.

Classifying Chemical Equations

In the next unit, we will examine a classification scheme based on chemical reactivity. Now, we will discuss another classification scheme that is based on the form of the chemical equation. The four categories are

a) single displacement reactions,
b) double displacement reactions,
c) synthesis (combination) reactions, and
d) decomposition reactions.

These classifications do not tell us why something happened, but they do tell us something about how many elements or compounds are involved.

Single Displacement Reactions

These reactions involve one pure element that reacts with a compound to become part of the compound; as a result one of the elements that was part of the compound is left behind as a pure element. An example is

$$Na + AuCl \rightarrow Au + NaCl$$

Many people think this type of reaction is similar to the situation when a couple is dancing and someone "cuts in."

Double Displacement Reactions

These reactions involve two compounds. They react with each other so that both partners are exchanged. This reaction can be compared with a situation where two couples are square dancing. An example is

$$NaCl + AgNO_3 \rightarrow NaNO_3 + AgCl$$

Decomposition Reactions

During decomposition reactions, a compound is broken down into two or more smaller pieces. An example is

$$CaCO_3 \rightarrow CO_2 + CaO$$

Combination Reactions

These reactions involve two or more chemicals that combine to form a single larger chemical. It is easy to see that combination reactions are the opposite of decomposition reactions. An example is

$$2\,Mg + O_2 \rightarrow 2\,MgO$$

Pre-workshop Problems

1.   For each compound, write the chemical formula and list how many of each type of atom are in each formula unit.

a. calcium phosphate

b. sodium nitrate

c. ammonium carbonate

d. iron (III) hydroxide

e. magnesium chlorate

2.   Using your textbook, give examples of each of the following types of unbalanced chemical equations. Make sure your answers are different than the examples shown in this unit.

a. single displacement

b. double displacement

c. synthesis (combination)

d. decomposition

Cooperative Group Problems

1. What is the difference between a coefficient and a subscript in a chemical equation? What is the result if you change a coefficient in an equation? What is the result if you change a subscript?

2. Consider the balanced chemical equation

$3 \, NaOH(aq) \; + \; H_3PO_4(aq) \; \rightarrow \; 3 \; H_2O(l) \; + \; Na_3PO_4(aq).$

a. Draw pictures of the reactants and products involved in this reaction.

b. Count the number of atoms of each type on the reactant and product sides of the equation.

Element	Reactant atoms	Product atoms
H		
O		
Na		
P		

c. Does this reaction obey the Law of Conservation of Matter? Explain.

3.   For each of the following reactions, write an unbalanced chemical equation and then balance it. For each step in the process, tell your group what you do and why you do it.

If you know the states of matter for the reactants and products, indicate them in your final answer.

a. Calcium hydroxide reacts with nitric acid to form calcium nitrate and water.

b. When heated, sodium peroxide will release oxygen gas and be transformed into sodium oxide.

c. Iron (III) oxide will react with carbon dioxide from the air to form iron (III) carbonate.

d. A blue-green solution of cobalt (III) chloride will turn colorless after zinc metal is added to the solution. A precipitate of cobalt metal grows as the blue-green color disappears from the solution.

4. For each reaction that you balanced in Problem 2, describe the type of reaction and explain how you made your decision.

a.

b.

c.

d.

5. Consider the four types of chemical equations discussed in this unit, and write a general form of an equation for each one. Where it is appropriate, use the general symbols A and B for metal atoms or ions and the general symbols C and D for nonmetal atoms or ions.

a. Single displacement

b. Double displacement

c. Synthesis

d. Decomposition

6.   For each of the following sentences, write the unbalanced chemical equation that summarizes the process and then balance the equation.

a. Lead (IV) oxide reacts with perchloric acid to form lead (IV) perchlorate and water.

b. Butyl alcohol (C_4H_9OH) is burned in air to form a hot gaseous mixture of carbon dioxide and water.

c. Platinum (IV) nitrate reacts with sodium bromide to form platinum (II) nitrate, bromine, and sodium nitrate.

Practice Exam Questions
Unit 5

1. What will be the coefficient of $O_{2(g)}$ when the following equation is balanced?

$$C_2H_5OH_{(l)} + O_{2(g)} \rightarrow CO_{2(g)} + H_2O_{(g)}$$

    a. 1
    b. 2
    c. 4
    d. 3
    e. 6

2. What will be the coefficient of $NO_{(g)}$ when the following equation is balanced?

$$NH_{3(g)} + O_{2(g)} \rightarrow NO_{(g)} + H_2O_{(g)}$$

    a. 2
    b. 4
    c. 5
    d. 6
    e. 10

3. When a solid piece of magnesium metal reacts with steam, hydrogen gas is given off and insoluble magnesium hydroxide in water forms. Write the unbalanced chemical equation that best represents this process.

    a. $Mg_{2(s)} + H_2O_{(l)} \rightarrow H_{(g)} + Mg(OH)_{2(s)}$
    b. $Mg_{(s)} + H_2O_{(g)} \rightarrow H_{(g)} + Mg(OH)_{2(s)}$
    c. $Mg_{(s)} + H_2O_{(l)} \rightarrow H_{2(g)} + Mg(OH)_{2(aq)}$
    d. $Mg_{(s)} + H_2O_{(l)} \rightarrow H_{2(g)} + Mg(OH)_{2(s)}$
    e. $Mg_{(s)} + H_2O_{(g)} \rightarrow H_{2(g)} + Mg(OH)_{2(s)}$

4. When liquid nitric acid decomposes, it leaves nitrogen dioxide gas, liquid water, and oxygen gas. Which of the following unbalanced chemical equations best represents this process?

    a. $HNO_{3(g)} \rightarrow NO_{(g)} + H_2O_{(l)} + O_{2(g)},$
    b. $HNO_{3(l)} \rightarrow NO_{2(g)} + H_2O_{(l)} + O_{2(g)}$
    c. $NO_{3(l)} \rightarrow NO_{2(g)} + H_2O_{(l)} + O_{2(g)}$
    d. $HNO_{3(g)} \rightarrow NO_{2(l)} + H_2O_{(g)} + O_{2(g)}$
    e. $HNO_{3(l)} \rightarrow NO_{2(l)} + H_2O_{(g)} + O_{2(l)}$

5. What will be the coefficient of $Si_3N_{4(s)}$ when the following reaction is balanced?

$$SiH_{4(g)} + NH_{3(g)} \rightarrow Si_3N_{4(s)} + H_{2(g)}$$

    a. 1
    b. 3
    c. 4
    d. 12
    e. 24

6. When the hydrochloric acid of your stomach reacts with sodium carbonate in water, it forms a sodium chloride solution, water, and carbon dioxide. Which of the following unbalanced chemical equations best represents this process?

    a. $HCl_{(aq)} + Na_2CO_{3(aq)} \rightarrow NaCl_{(s)} + H_2O_{(s)} + CO_{2(g)}$
    b. $HCl_{(s)} + Na_2CO_{3(s)} \rightarrow NaCl_{(s)} + H_2O_{(s)} + CO_{2(g)}$
    c. $HCl_{(aq)} + Na_2CO_{3(aq)} \rightarrow NaCl_{(aq)} + H_2O_{(l)} + CO_{2(g)}$
    d. $HCl_{(l)} + Na_2CO_{3(s)} \rightarrow NaCl_{(s)} + H_2O_{(g)} + CO_{2(g)}$
    e. $HCl_{(aq)} + Na_2CO_{3(s)} \rightarrow NaCl_{(aq)} + H_2O_{(s)} + CO_{2(g)}$

7. What will be the coefficient of $I_{2(s)}$ when the following equation is balanced?

$$CuSO_{4(aq)} + KI_{(s)} \rightarrow CuI_{(s)} + I_{2(s)} + K_2SO_{4(aq)}.$$

    a. 1
    b. 2
    c. 4
    d. 5
    e. 8

8. What will be the coefficient of $Fe(NO_3)_{3(aq)}$ when the following equation is balanced?

$$Fe_2O_{3(s)} + HNO_{3(aq)} \rightarrow Fe(NO_3)_{3(aq)} + H_2O_{(l)}.$$

    a. 1
    b. 2
    c. 3
    d. 4
    e. 6

9. When dihydrogen sulfide gas burns in the air, it produces sulfur dioxide gas and water vapor. Which of the following best describes this process?

      a. $H_2S_{(l)} + O_{2(g)} \rightarrow SO_{(g)} + H_2O_{(l)}$
      b. $H_2S_{(g)} + O_{2(g)} \rightarrow SO_{2(g)} + H_2O_{(l)}$
      c. $H_2S_{(g)} + O_{2(l)} \rightarrow SO_{2(l)} + H_2O_{(g)}$
      d. $H_2S_{(l)} + O_{2(g)} \rightarrow SO_{2(l)} + H_2O_{(g)}$
      e. $H_2S_{(g)} + O_{2(g)} \rightarrow SO_{2(g)} + H_2O_{(g)}$

10. When solid calcium carbonate is heated strongly, carbon dioxide gas is given off and a residue of calcium oxide is left over. Which of the following best describes this process?

      a. $CaCO_{3(g)} + heat \rightarrow CO_{2(g)} + CaO_{(s)}$
      b. $CaCO_{3(l)} + heat \rightarrow CO_{2(l)} + CaO_{(l)}$
      c. $CaCO_{3(s)} + heat \rightarrow CO_{2(g)} + CaO_{(s)}$
      d. $CaCO_{3(s)} + heat \rightarrow CO_{2(s)} + CaO_{(g)}$
      e. $CaCO_{3(s)} + heat \rightarrow CO_{2(l)} + CaO_{(l)}$

Unit 6: Predicting the Results

This unit deals with a critical concept in chemistry: predicting the course of chemical reactions. At any time on Earth, there are literally millions of chemical reactions occurring simultaneously. To understand as many of these reactions as possible, general types of reactions have been recognized and described. Classifying types of reactions dramatically reduces the amount of memorization that we need to do to understand them. You will have to memorize many facts to become good at chemistry, but understanding general principles will lessen the load.

Types of Reactions Based on Chemical Properties and Reactivity

The millions of chemical reactions can be classified into thousands of different general types of reactions. In Unit 5, we classified chemical reactions based on the *form* of the equation; the classifications included single-displacement, double displacement, synthesis, and decomposition reactions. At this point, we will only study a few of the most important reaction types, and will try to explain why some chemicals react as they do. Three major *reactivity-based* categories for reactions are acid/base, precipitation, and oxidation/reduction. Chemists use these categories to understand and predict chemical reactions.

Reaction Tendencies

If you observe society, you will see that each person tends to react predictably in certain ways. People with short tempers are generally more likely to get into arguments than people who are more easygoing. Likewise, chemicals sometimes seem to have personalities because they predictably follow the same patterns of behavior when involved in reactions. If you wish to predict what a chemical will do in a particular reaction, it is useful to know how it normally reacts. As the reaction types are described in this unit, the compounds that participate in each type of reaction will also be described. Make sure that you learn how to identify each type of compound and predict what it might do.

Acid/Base Reactions

There are many definitions for acids and bases, but we will only use one definition for acids in this workbook, the Brönsted-Lowry definition. This definition says that an acid is a chemical that will transfer a H^+ ion to another chemical. The chemical that receives the H^+ is called a base. If you are given an equation for an acid/base reaction, you can easily recognize that it is an acid/base reaction; simply apply these definitions. This is illustrated in Figure 6.1.

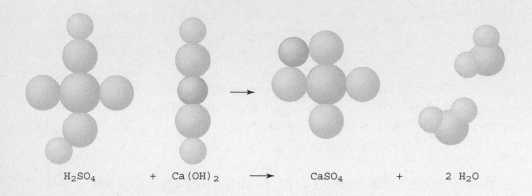

$$H_2SO_4 \quad + \quad Ca(OH)_2 \quad \longrightarrow \quad CaSO_4 \quad + \quad 2\ H_2O$$

Figure 6.1 An acid-base reaction.

In this case, H^+ is transferred from the sulfuric acid to the hydroxide to form water. We can write a partial reaction that describes only the H^+ transfer to the hydroxide:

$$H^+ + OH^- \rightarrow H_2O$$

Although some acid/base reactions form water, not all of them will. Here is an example of another acid/base reaction. The H^+ is transferred from the hydrochloric acid to the ammonia molecule to form the ammonium ion:

$$NH_3 + HCl \rightarrow NH_4Cl$$

Again, a partial reaction makes this hydrogen transfer to the ammonia molecule more obvious:

$$NH_3 + H^+ \rightarrow NH_4^+$$

Acids

In our definition of an acid, we specified that it needs to have a hydrogen atom that can be transferred as a hydrogen cation to the base. Since we have studied nomenclature, we are already familiar with most of the common acids. They were composed of a stable anion combined with a hydrogen cation or multiple hydrogen cations as needed to balance the charge. An acid is capable of transferring a hydrogen cation to a base.

Bases

A very common type of base is a metal hydroxide. Since the O and H are non-metals, the metal hydroxide is ionic. Thus, there is a negative charge on the OH group, which is why we call it a hydrox**ide**. Because of the negative charge, an OH- group is very good at attracting a H^+:

$$H^+ + OH^- \rightarrow H_2O$$

If an OH group is combined with non-metals, then it will not have a negative charge or act as a base. An OH group is not termed a hydroxide unless it is in an ionic compound.

Most other negatively charged ions, especially highly charged ones, can also accept the positively charged H^+. One important example is carbonate. Carbonate can accept two hydrogen ions to form carbonic acid, which subsequently decomposes into carbon dioxide and water. The basicity of carbonate is very important for antacids, geological sciences, and the stability of our blood:

$$CO_3^{2-}{}_{(aq)} + 2\,H^+{}_{(aq)} \rightarrow H_2CO_{3(aq)} \rightarrow H_2O_{(l)} + CO_{2(g)}$$

Another very important base is ammonia, which was shown earlier in this unit. It is involved in the preparation of fertilizers and important specialty chemicals. Ammonia reacts with acids to form the ammonium ion.

Precipitation Reactions

A very popular demonstration of chemical reactions involves the precipitation of a solid from two clear, colorless liquids. A clear, colorless solution of lead (II) nitrate will react with a clear, colorless solution of sodium chloride to form a yellow solid, which seems to appear from nowhere. This type of reaction looks like magic to people who do not normally watch chemistry in action. However, the key to understanding these reactions is simply an understanding of solubility. Some compounds dissolve in water, but others do not. It is possible to mix two soluble compounds that are dissolved in water such that an insoluble solid precipitates from the reaction mixture. If we wish to accurately predict the course of these reactions, we first need to learn which compounds are soluble and which are not.

Solubility Rules for Ionic Compounds

Most textbooks have a list of rules for you to memorize; this enables you to predict which ionic compounds dissolve in water and which do not. Similarly, the flash card packet that accompanies this workbook includes a set of solubility rules that generalizes the patterns of solubility for ionic substances. There are a few important points about this set of generalizations and others. First, the set of rules is far from complete. To keep the list of rules manageable, only the most common ions are included. Second, the criterion for deciding what is soluble is somewhat arbitrary. No compound is completely insoluble; they all dissolve at least a little bit in water. A good set of solubility rules will clearly state a quantitative definition of solubility that was used to write the rules. The criterion must include the *amount of compound* that will dissolve in a *particular amount of water* at a *particular temperature*.

Predicting Precipitation Reactions

To make accurate predictions about precipitation reactions, you must first have an understanding of the solubility guidelines. Many people find that learning the solubility rules is much harder than applying them. If a pair of ions that form an insoluble compound is brought together in water, then the insoluble compound will precipitate. Here is an example of a precipitation reaction:

$$AgNO_3(aq) + NaI(aq) \rightarrow AgI(s) + NaNO_3(aq)$$

Silver nitrate, like all nitrates, is soluble. Sodium iodide, like most iodides, is also soluble. Thus, they both form homogeneous solutions in water. Silver iodide is one of the few insoluble iodides, so when a solution containing the silver ion is mixed with a solution containing the iodide ion, solid silver iodide forms. When you are trying to predict chemical reactivity, knowing which ions generally precipitate is just as important as knowing which compounds are acids and how to name compounds.

Oxidation/Reduction Reactions

Predicting the course of oxidation/reduction reactions is a skill that is quite advanced. Thus, at this stage, we will only concern ourselves with the identification of oxidation/reduction reactions.

Historically, oxidation referred only to the reaction of a substance with oxygen to form an oxide. Many substances will participate in this type of reaction. For instance, sulfur will react with oxygen to form either sulfur dioxide or sulfur trioxide.

Later, scientists recognized that most compounds that reacted with oxygen would also react in a similar fashion with chlorine or fluorine. Sulfur can react to form sulfur tetrafluoride or sulfur hexafluoride. In either case, our view is that the sulfur transfers electrons to oxygen to form oxides or to fluorine to form fluorides. When studying acid/base reactions, we needed to keep track of H^+ transfers. When studying oxidation/reduction reactions, we will need to keep track of electron transfers.

Definitions: Oxidation and Reduction

An oxidation/reduction reaction is one in which electrons are transferred. In order to have an oxidation/reduction reaction, you must have an element that loses electrons and an element that gains them. If an element loses electrons, it is oxidized. If an element gains electrons, it is reduced. (Many students can remember the mnemonic, "LEO says GER.". This mnemonic stands for Loss of Electrons is Oxidation and Gain of Electrons is Reduction. This relationship is similar to a money transfer. If you give one of you students a dollar for answering a question well, you have a money transfer. (You lost money and the student gained money.) Any transfer involves both a loss and a gain.

Definitions: Oxidizing Agents and Reducing Agents

If you wanted to clean the floor, you would use some sort of cleaning agent, like soap or detergent. The role of the cleaning agent is to remove dirt from the floor. During the transfer of dirt from the floor to the cleaning agent, the floor becomes clean and the cleaning agent becomes dirty, as shown by the following equation:

dirty floor + cleaning agent (soap) → clean floor + dirty soap

The terminology for oxidizing agents is similar. If you have iron metal and want iron (III) ions, you must remove electrons from the metal (oxidize the metal). Anything that could take electrons from iron metal would allow you to accomplish the oxidation of the iron, just like anything that could take the dirt off the floor would accomplish the mission of cleaning the floor. Whatever oxidizes the iron gains electrons, and therefore becomes reduced. During the process of oxidizing something, the oxidizing agent becomes reduced, just as during the process of cleaning something, the cleaning agent becomes dirty:

$$Fe + \text{oxidizing agent} \rightarrow Fe^{3+}$$
Iron is oxidized by the oxidizing agent.

Fluorine would be able to oxidize iron metal according to the following equation:

$$2\,Fe + 3\,F_2 \rightarrow 2\,FeF_3$$

During the process, fluorine is reduced by iron, the reducing agent:

$$F_2 + \text{reducing agent} \rightarrow 2\ F^-$$

Oxidation Numbers

Oxidation numbers were invented so that chemists could track electron transfer reactions. For compounds that are not ionic, like carbon dioxide, it does not really make sense to talk about the charge on the carbon atom. Carbon dioxide stays intact and never dissociates into carbon ions and oxygen ions. Even so, it is useful to think of the formation of carbon dioxide as involving a transfer of electrons from carbon to oxygen. The most important rule when determining oxidation numbers is that the sum of the oxidation numbers in a molecule or ion must equal the charge on the molecule or ion. For simple ionic compounds like sodium oxide, the oxidation number of each element equals the charge of its ion. Sodium has an oxidation number of +1 and oxygen has an oxidation number of –2. You can look in your textbook for a complete list of rules for determining oxidation numbers.

Change in Oxidation Numbers

The reaction between potassium permanganate and hydroiodic acid to form manganese (IV) oxide and iodine is an example of an oxidation/reduction reaction. The way to tell that it is an oxidation/reduction reaction is to determine the oxidation numbers of all the elements in the reactants and products of the reaction. If the oxidation numbers of any elements change during the reaction, then you know it is an oxidation/reduction reaction.

$$2\ KMnO_4 + 6\ HI \rightarrow 2\ MnO_2 + 3\ I_2 + K_2O + 3\ H_2O$$

Element Symbol	Reactant Oxidation Number	Product Oxidation Number	Half Reaction
K	+1	+1	None, not involved
Mn	+7	+4	$Mn^{7+} + 3e^- \rightarrow Mn^{4+}$
O	-2	-2	None, not involved
H	+1	+1	None, not involved
I	-1	0	$I^- \rightarrow 1/2\ I_2 + e^-$

Naming the Oxidizing Agent

There are two ways that are commonly used to identify the oxidizing and reducing agents in a reaction. In one, we identify the atom that undergoes the change in oxidation number. In the other, we identify the compound containing the atom that undergoes the change in oxidation number. In the above reaction, potassium permanganate contains the manganese atom, which went from having an oxidation number of +7 to +4. That means it was reduced (gained electrons) and was the oxidizing agent. Some chemists would say that the element manganese was the oxidizing agent, but others would say that the compound potassium permanganate was the oxidizing agent. Both are correct, but they use different conventions.

Pre-workshop Problems

1.   For each of the following compounds or ions, determine the oxidation number of each element:

a. sodium oxide

b. sodium peroxide

c. aluminum nitrate

d. carbonate

e. zinc phosphite

f. oxygen difluoride

g. ammonium chromate

2. Determine whether each compound is an acid, a base, or neither. Write the name of each compound.

a. CaS

b. $NaNO_3$

c. $HC_2H_3O_2$

d. HBr

e. $Ca(OH)_2$

f. CH_3OH

3.   Using the rules of solubility, predict whether each of the following compounds are soluble or not. Write the formula for each compound.

a. calcium sulfide

b. sodium nitrate

c. lead (II) chloride

d. ammonium phosphate

e. barium sulfate

f. iron (III) hydroxide

Cooperative Group Problems

1.   For each of the following reactions, decide whether it is an acid/base reaction, a precipitation reaction, or an oxidation/reduction reaction. If it is an oxidation/reduction reaction, state which reactants are the oxidizing and reducing agents.

a.   Sodium hydride reacts with silver nitrate to form silver metal, hydrogen, and sodium nitrate.

b.   Sodium sulfide reacts with hydrochloric acid to form hydrogen sulfide and sodium chloride.

c.   Sodium sulfate reacts with barium chloride to form sodium chloride and barium sulfate.

d.   Sodium hydroxide reacts with hydrobromic acid to form water and sodium bromide.

e.   Sodium chloride reacts with mercury (I) nitrate to form sodium nitrate and mercury (I) chloride.

f.   Mercury (I) nitrate reacts with zinc to form mercury and zinc nitrate.

2. Using the solubility flash cards, play the "Around the World" game with your team. For each compound, decide whether it is soluble or insoluble, and decide whether it follows a rule or is an important exception to a rule. These flash cards are based entirely on anion rules, which is important in deciding whether a compound is soluble by rule or by exception. If your instructor prefers to define the rule, "All group I salts are soluble," then you should change the flash cards appropriately before you start playing.

3. For each of the following reactions, identify the type of reaction, predict the products, and write a balanced equation that describes the reaction.

a. Barium hydroxide reacts with chlorous acid.

b. Lead (II) chloride reacts with sodium sulfate.

c. Sodium hydride reacts with hydrochloric acid.

d. Silver nitrate reacts with iron (II) bromide.

e. Sodium carbonate reacts with acetic acid.

f. Sodium carbonate reacts with calcium iodate.

Practice Exam Questions
Unit 6

1. A precipitation reaction is a reaction that forms

    a. a gas
    b. a solid
    c. a liquid
    d. an aqueous solution
    e. a mixture

2. A strong electrolyte is a substance that, when dissolved in water, produces a solution that conducts electricity

    a. not at all
    b. very poorly
    c. very well
    d. slightly
    e. none of the above

3. A solid that will dissolve completely in water is called a

    a. strong electrolyte
    b. weak electrolyte
    c. insoluble solid
    d. slightly soluble solid
    e. soluble solid

4. Which is not soluble in water?

    a. KNO_3
    b. $NaCl$
    c. NH_4I
    d. KOH
    e. $BaSO_4$

5. Which is soluble in water?

    a. CaS
    b. $Fe(OH)_2$
    c. $BaCO_3$
    d. Na_3PO_4
    e. $Zn_3(PO_4)_2$

6. The complete ionic equation shows

    a. only the participating ions
    b. the complete formulas of the reactants and products as neutral species
    c. the formulas of the reactants but not the products
    d. the forms of the reactants and products in the reaction
    e. the spectator ions

7. The ions that directly participate in the reaction are shown in

    a. the molecular equation
    b. the complete ionic equation
    c. the spectator ion equation
    d. the net ionic equation
    e. none of the above

8. The following equation is an example of what type of reaction?

$$Zn_{(s)} + 2\ HNO_{3(aq)} \rightarrow Zn(NO_3)_{2(aq)} + H_{2(g)},$$

    a. acid-base
    b. oxidation-reduction
    c. precipitation reaction
    d. double-displacement reaction
    e. decomposition

9. The following equation is an example of what type of reaction?

$$Fe(OH)_{2(s)} + H_2SO_{4(aq)} \rightarrow FeSO_{4(aq)} + H_2O_{(l)}.$$

    a. acid-base
    b. oxidation-reduction
    c. precipitation reaction
    d. single displacement reaction
    e. combination reaction

10. During a synthesis reaction (also called a combination reaction),

    a. a compound is formed from simpler materials
    b. a compound is broken down into smaller compounds
    c. an acid-base reaction occurs
    d. a single displacement reaction occurs
    e. a double displacement reaction occurs

11. Which of the following best describes the following equation?

$$2\,Mg_{(s)} + O_{2(g)} \rightarrow 2\,MgO_{(s)}$$

a. acid-base reaction
b. combustion reaction
c. double displacement reaction
d. precipitation reaction
e. single displacement reaction

12. When aqueous potassium bromide and aqueous silver nitrate react, the precipitate they produce will be

      a. KNO_3
      b. $AgBr$
      c. $AgNO_3$
      d. KBr

13. In the reaction between aluminum and bromine, how many electrons do 2 atoms of aluminum lose?

$$2Al_{(s)} + 3Br_{2(l)} \rightarrow 2AlBr_{3(s)}$$

      a. 2
      b. 3
      c. 4
      d. 6
      e. 12

14. Which of the following pairs of equations summarizes the flow of electrons for the following reaction?

$$2Mg_{(s)} + O_{2(g)} \rightarrow 2MgO_{(s)}.$$

a. $Mg^{2+} + 2e^- \rightarrow Mg$; $O^{2-} + 2e^- \rightarrow O$
b. $Mg \rightarrow Mg^{2+} + 2e^-$; $O^{2-} + 2e^- \rightarrow O$
c. $Mg \rightarrow Mg^{2+} + 2e^-$; $O + 2e^- \rightarrow O^{2-}$
d. $Mg^{2+} + 2e^- \rightarrow Mg$; $O + 2e^- \rightarrow O^{2-}$
e. none of the above

Unit 7: The Art of Counting without Counting

The chemical equations that we have discussed are written in terms of the numbers of molecules that react. When you have a jar of sugar, there is no easy way to directly determine how many sugar molecules are in that jar. On the other hand, it is very easy to determine the mass of the sugar in the jar by using a balance. Chemists recognized the need for a conversion factor that would convert the mass of a substance into the number of molecules of that substance.

Molar Mass of the Elements

The conversion factor that relates a mass of a substance to the number of molecules it contains is called molar mass, and it is expressed in grams/molecule. The molar mass of each element is listed on the periodic table. For instance, the molar mass of chlorine atoms is 35.45 g/mol. This means that 35.45 g of chlorine equals 1 mole of chlorine atoms. Writing this equality out gives

$$35.45 \text{ g Cl atoms} = 1 \text{ mol Cl atoms}$$

Now that we know the equality, we can easily build two conversion factors to convert between grams and moles of Cl:

$$\frac{35.45\text{g}}{1\text{mol}} \text{ and } \frac{1\text{mol}}{35.45\text{g}}$$

The Mole

In concept, the term "mole" is very similar to the term "dozen". For example, one dozen equals 12 units. Thus, one dozen eggs equals 12 eggs, and one dozen chlorine atoms equals 12 chlorine atoms. If you have 2.5 dozen chlorine atoms, you can easily determine the number of chlorine atoms through the following calculation:

$$2.5 \text{ dozen Cl atoms x } \frac{12 \text{ Cl atoms}}{\text{dozen Cl atoms}} = 30 \text{ Cl atoms}$$

To convert moles to individual units, you will use the same calculation with one slight difference: you must consider that one mole is a lot bigger than one dozen. In fact, one mole is 6.02×10^{23} units (Avogadro's number). This number is almost unfathomably large.   Making this change to the equation, you can determine that 2.5 moles of chlorine atoms equals

$$2.5 \text{ mol Cl atoms} \times \underline{\frac{6.02 \times 10^{23} \text{ Cl atoms}}{\text{mol Cl atoms}}} = 2.1 \times 10^{24} \text{ Cl atoms}$$

If you can do this dimensional analysis with ease, you can easily track the numbers of molecules, atoms, and ions in a given amount of substance.

Molar Mass of Compounds

Compounds are composed of atoms. The molar mass of a compound is a sum of the molar masses of the atoms that compose the compound. The molar mass of glucose ($C_6H_{12}O_6$) can be calculated by adding 6 times the molar mass of C, 12 times the molar mass of hydrogen, and 6 times the molar mass of O:

$$\left(6 \times 12.01 \frac{g}{mol} \right) + \left(12 \times 1.008 \frac{g}{mol} \right) + \left(6 \times 16.00 \frac{g}{mol} \right) = 180.2 \frac{g}{mol}$$

This information allows us to calculate the number of molecules in a given mass of glucose. If we have 15.0g of glucose, we can use the molar mass of glucose to find out the number of molecules. To derive the number of grams from the number of moles of a substance, we use the following equation:

$$15.0\text{g glucose} \times \underline{\frac{1 \text{ mol glucose}}{180.2 \text{ g glucose}}} = 0.0832 \text{ mol glucose}$$

To calculate the number of molecules of glucose, we simply use Avogadro's number:

$$0.0832 \text{ mol glucose} \times \underline{\frac{6.02 \times 10^{23} \text{ molecules glucose}}{\text{mol glucose}}} =$$

$$5.01 \times 10^{22} \text{ molecules glucose}$$

Formulas

At times, chemists need to know how many atoms of a particular type are in a given amount of a substance. Formulas tell us how many atoms of each type are in a molecule. Because we know that the formula of glucose is $C_6H_{12}O_6$, we can write the following equality:

$$12 \text{ H atoms} = 1 \text{ glucose molecule}$$

Although a glucose molecule contains components other than the 12 H atoms, the preceding equation helps us to understand the relationship between the number of glucose molecules and the number of hydrogen atoms they contain. Continuing to evaluate our example of 15.0g of glucose, we can calculate the number of hydrogen atoms that sample contains:

$$5.01 \times 10^{22} \text{ glucose molecules} \times \frac{12 \text{ H atoms}}{\text{molecule glucose}} = 6.01 \times 10^{23} \text{ H atoms}$$

Percent Composition

In addition to using formulas, we can use another method to describe the composition of a substance; this method is called the percent composition of a substance. This is another concept that you have already learned in a slightly different context. For example, consider a class with 10 men and 15 women; you can probably calculate the percentage of students that are women. The fraction of women is the number of women divided by the total number of people in the class. To change the fraction into a percentage, multiply by 100. Thus, we have

$$\frac{15 \text{ women}}{(10 + 15) \text{ people}} \times 100 = 60.0\% \text{ women.}$$

Percent Composition by Mass

In the preceding example, we were concerned with the number of women in the class, so we used the number of women in our calculation. For molecules, we find it easier to measure the mass than to measure the number of molecules. For that reason, chemists generally calculate and use the percent composition by mass to describe the constituents of a molecule or mixture. We simply replace the number of items in the formula with the mass of the items.

To help us calculate the percent composition of glucose by mass, we will simply place the information about the components of glucose ($C_6H_{12}O_6$) into a table. The fraction of each element is calculated in the right column. As previously determined, the total molar mass of glucose is 180.2 g/mol.

Element	Number in Formula	Atomic Mass	Total Mass	Fraction by mass of element in glucose
C	6	12.01	72.06	72.06/180.2=0.400
H	12	1.0079	12.096	12.096/180.2=0.067
O	6	16.00	96.00	96.00/180.2=0.533

Thus, the percent composition by mass of glucose is 40.0% C, 6.7% H, and 53.3% O. It is important that the sum of these percentages equals 100.

Determining Formulas

When a new compound is made for the first time, chemists will send it for an elemental analysis so they can determine the formula of the compound. The lab that does the elemental analysis will determine the percent composition by mass for the compound. A little poem will help us remember the steps that allow us to convert the percent composition data into a formula:

% to mass
mass to mol
divide by small
multiply until whole

The percent composition of a compound is not dependent on the amount of compound. If we assume that we have 100g of our compound, then the percentages for each element can be changed directly to masses. Since formulas tell us the ratios of the numbers of each atom, we must change masses to numbers of moles using the molar masses for each atom. At this stage, we can write a formula with the proper ratio, but we would rather write the formula using whole numbers. The last two steps accomplish this goal.

Empirical Formula Determination

Nitroglycerine is composed of carbon (15.88%), hydrogen (2.22%), oxygen (63.42%), and nitrogen (18.50%). To determine its empirical formula, we must take several steps. First, we must assume that we have 100 g of compound. If we have 100 g of nitroglycerine, the sample would contain 15.88 g C, 2.22 g H, 63.42 g O, and 18.50 g N. Converting these masses to numbers of moles is done as follows:

$$15.88 \text{ g C} \times \frac{1 \text{mol C}}{12.01 \text{ g C}} = 1.322 \text{ mol C,}$$

$$2.22 \text{ g H} \times \frac{1 \text{ mol H}}{1.0079 \text{ g H}} = 2.203 \text{ mol H,}$$

$$63.42 \text{ g O} \times \frac{1 \text{ mol O}}{16.00 \text{ g O}} = 3.964 \text{ mol O,}$$

$$18.50 \text{ g N} \times \frac{1 \text{ mol N}}{14.01 \text{ g N}} = 1.320 \text{ mol N.}$$

Writing a formula from this data gives us

$$C_{1.322}H_{2.203}O_{3.964}N_{1.32}.$$

Dividing each subscript by the lowest of the four subscripts gives us

$$C_{1.322/1.32}H_{2.203/1.32}O_{3.964/1.32}N_{1.32/1.32} = C_{1.00}H_{1.667}O_{3.00}N_1.$$

To eliminate the final non-whole number, we must multiply all subscripts by the same number; but we must be careful to preserve the ratio between atoms. In this case, we will multiply all four subscripts by a factor of 3. Three is the correct factor because the subscript for H is 1.667, which is easy to recognize as 1 2/3. (Recognizing the fractional equivalents of decimal numbers makes it easy to eliminate them.) Multiplying 1 2/3 by 3 yields a whole number: 5. After multiplying all subscripts by 3, we have our final empirical formula:

$$C_3H_5O_9N_3$$

Empirical vs. Molecular Formulas

It is important to recognize that when we convert percent composition data to a formula, we get an empirical formula, not necessarily a molecular formula. We will investigate this phenomenon in the cooperative group exercises.

Pre-workshop Problems

1.   For each of the following compounds, calculate the molecular mass:

a. magnesium acetate

b. iron (III) sulfate

c. methyl alcohol

 2.   For each compound in Question 1, calculate how many molecules would be present in 1.00 g of the compound.

a.

b.

c.

3.   Pretend you just won the Ultra-Lotto grand prize of 1 mol of dollar bills. In order to count your money, you hire everyone in the United States to help you count.   Assuming that each person could count one dollar per second, how long would it take to count your winnings?   Just take a guess, then figure it out using the following steps:

a. How many dollars would each person have to count?

 b. How many dollar bills could each person count in a year, assuming they never took a break?

c. How many years would it take to count all of your money?

Cooperative Group Problems

1. Ethyne and benzene are two very common organic compounds.

a. For each of these compounds, calculate a percent composition.

b. Compare the results of both calculations. Can you describe and explain the relationship between your two answers?

c. Using your textbook, draw the structures of these two molecules.

d. When you calculate a formula from a percent composition, you have an empirical formula. What other information do you need to get the molecular formula?

2. A biologist analyzes a dipeptide and finds the percent composition of the compound to be 32.7% C, 4.8% H, 19.0% N, and 43.5% O. What is the formula of the dipeptide?

3. A hydrocarbon (compound containing only hydrogen and carbon) is 85.7% carbon and has a molecular mass of 84.15 g/mol.

a. What is the empirical formula of the compound?

b. What is the formula of the compound?

Practice Exam Questions
Unit 7

1. Which represents the smallest number of atoms?

   a. 25.0 g Be
   b. 25.0 g O
   c. 25.0 g Cl
   d. 25.0 g Ar
   e. all the same

2. Which represents the greatest number of atoms?

   a. 45.0 g Na
   b. 45.0 g Mg
   c. 45.0 g P
   d. 45.0 g F
   e. all the same

3. How many atoms are present in 100 g of nitrogen?

   a. 7.14
   b. 4.3×10^{23}
   c. 8.4×10^{26}
   d. 1.2×10^{-23}
   e. none of the above

4. How many atoms are present in 100 g of xenon?

   a. 0.76
   b. 7.91×10^{27}
   c. 1.26×10^{-24}
   d. 4.59×10^{23}
   e. none of the above

5. What is the molar mass of ethanol, CH_3CH_2OH?

   a. 34.0 g/mol
   b. 45.0 g/mol
   c. 46.0 g/mol
   d. 123 g/mol
   e. none of the above

6. 0.512 mol of an element has a mass of 16.42 g. Identify the element.

    a. Si
    b. P
    c. S
    d. Cl
    e. Ar

7. How many oxygen atoms are contained in 5.02 g of iron (III) sulfate (MW = 399.9 g/mol)?

    a. 0.0126
    b. 0.151
    c. 9.07×10^{22}
    d. 7.56×10^{21}
    e. 2.26×10^{22}

8. How many molecules are in 6.71g of ethanol, CH_3CH_2OH?

    a. 8.78×10^{22}
    b. 0.146
    c. 7.21×10^{22}
    d. 1.12×10^{23}
    e. 2.24×10^{23}

9. What is the empirical formula for a compound that is composed of 11.66 g of iron and 5.01 g of oxygen?

    a. Fe_2O_3
    b. FeO
    c. Fe_3O_2
    d. FeO_3
    e. Fe_3O_4

10. What is the empirical formula of a compound that has 58.84% barium, 13.74% sulfur, and 27.42% oxygen?

    a. BaSO
    b. $BaSO_3$
    c. Ba_2S_2O
    d. $BaSO_4$
    e. $BaSO_2$

11. What is the empirical formula of a compound that has 32.13% of aluminum and 67.87% of fluorine?

    a. AlF
    b. Al_2F
    c. Al_3F
    d. Al_2F_3
    e. AlF_3

12. What is the empirical formula of a 0.5998 g sample of a compound that contains 0.2322 g of carbon, 0.05848 g of hydrogen, and 0.3091 g of oxygen?

    a. CHO
    b. $C_2H_3O_2$
    c. CH_3O
    d. C_3HO_3
    e. CHO_3

13. What is the molecular formula of a compound that has an empirical formula of CH_2O and a molecular mass of 180 g/mol?

    a. $C_9H_{18}O_8$
    b. $C_5H_{10}O_5$
    c. $C_8H_{20}O_4$
    d. $C_6H_{12}O_6$
    e. $C_8H_{16}O_8$

14. How many empirical units are in a compound that has an empirical mass of 14 g/mol and a molecular mass of 126 g/mol?

    a. 6
    b. 7
    c. 8
    d. 9
    e. 10

Unit 8: How Much Do I Get?

If you are running a chemical company or a research lab at a university, you might be very interested in knowing how much product you can make. This will depend on two things: how much starting material you have at your disposal and Mother Nature. A balanced chemical equation tells us the relative amount of product we might be able to get from our starting materials.

Non-Chemical Examples

If you have mastered the conversion that converts a compound's mass to its number of moles, then you already have all of the tools to solve even the most difficult stoichiometry problems. You do stoichiometry problems on a regular basis in real life. The goal of this unit is to increase your ability to recognize real-life stoichiometry problems and to show you explicitly how to solve them. Then, we will apply the method that you already use in real life to chemical stoichiometry problems.

Tennis Balls

An example of a non-chemical stoichiometry problem involves the packaging of tennis balls. If you have 54 tennis balls and 20 empty packages, how many packages of tennis balls can you sell? How many items of which kind will you have left? Solving this explicitly using a balanced equation should help us in the future with chemical problems.

Step 1: Write a balanced equation.

> 3 tennis balls + 1 empty package → 1 package for sale

Step 2: Answer the question, "How much product can I make with each reactant?"

> To answer this question, we need to conduct a dimensional analysis with the coefficients of the balanced equation. In real life, we would not have to write this problem because we solve it intuitively, but explicitly writing the steps will help us understand the process of solving the problem. Since it takes 3 tennis balls to fill 1 package, we can write

> 54 tennis balls x $\underline{\text{1 package for sale}}$ = 18 packages for sale.
>   3 tennis balls

Also, we write

20 empty packages x $\dfrac{\text{1 package for sale}}{\text{1 empty package}}$ = 20 packages for sale.

Step 3: Identify the reactant that allows us to create the fewest units of product as the limiting reagent.

In the preceding example, we can make 18 filled packages of balls for sale; then we run out of tennis balls and we stop production.

Step 4: Calculate the leftover reagent.

By using the number of units produced and dimensional analysis, we can easily figure out how much of the leftover reagent we have after production ceases.

18 full packages x $\dfrac{\text{1 empty package}}{\text{1 full package}}$ = 18 empty packages.

The resulting number of empty packages equals the number of beginning number of empty packages minus the number we used. That is, we are left with 20-18=2 empty packages.

An Alternate Method

Chemists often use a different method to determine the limiting reagent. Instead of asking how much product we can make from each starting material, we could ask how much of one starting material would we need to use all of the other starting material? A comparison of how much we have to how much we need allows us to determine the limiting reagent. We can use the preceding example to illustrate this method.

Alternate Step 2: Answer the question: "How many packages would I need to use all the tennis balls?"

To do this, we need to conduct a dimensional analysis with the coefficients of the balanced equation. Since it takes 3 tennis balls to fill 1 package, we can write

54 tennis balls x $\dfrac{\text{1 empty package}}{\text{3 tennis balls}}$ = 18 packages needed.

Since we have 20 packages, we have more than enough packages and will run out of tennis balls. We would then use the number of tennis balls to figure out how many full packages we could make.

Grams to Moles

Often, chemicals are measured on a balance. The coefficients of chemical equations refer to numbers of moles. Thus, in order to start the preceding procedure, we generally need to convert our formula from grams to moles. Later in the course, you will learn other ways to calculate the number of moles of a substance. Titration problems, gravimetric analysis problems, and ideal gas problems are other forms of stoichiometry problems.

Practice

If you understand how to convert grams to moles and you understand the step-by-step explanation of the method used in the tennis ball example, then you know what you need in order to solve stoichiometry problems. You will just need to practice applying the same methods you use to solve regular stoichiometry problems to solve chemistry problems.

Simpler Problems

Our tennis ball example is one of the most difficult types of stoichiometry problems, because you have to figure out which item you will run out of (the limiting reagent). A simpler question would be "How many packages of tennis balls can you sell if you have 90 tennis balls?" In this case, you assume that you have as many empty packages as you need to use all the tennis balls. To solve this problem, a single calculation is sufficient:

$$90 \text{ tennis balls} \times \frac{1 \text{ package}}{3 \text{ tennis balls}} = 30 \text{ packages.}$$

Percent Yield

Not every reaction occurs as planned. Sometimes reactants decompose and sometimes there are side reactions that give you products that you did not want. For example, you might lose some of the tennis balls that you wished to sell or spill some ink on them. Our calculations assumed that the packaging process went perfectly, so we would be able to sell 18 packages of tennis balls. That would be 100% yield. The calculation of percent yield is very much like the calculation of percent composition. The fractional yield is the actual yield from the reaction divided by the maximum possible yield. Multiplying the fractional yield by 100 gives the percent yield. If your employees had 54 tennis balls and 20 empty packages but only managed to fill 12 packages, then your percent yield is

$$\frac{12 \text{ packages for sale}}{18 \text{ packages for sale}} \times 100 = 66.7\%.$$

Pre-workshop Problems

1.   You need to build a water-powered toy car using 4 wheels, 1 chassis, 3 decals, and 5 ounces of water. You have 28 wheels, 15 chassis, 18 decals, and 2 quarts of water.

a. Write an equation representing the process of assembling a water-powered toy car using coefficients and one or two letter abbreviations. Make it look as much like a chemical equation as possible.

b. Figure out how many toy cars you can make and how many of each piece you have left over after all the cars are made.   You may find it helpful to review the procedure that was presented earlier in the unit.

c. Solve this problem using the alternative method for identifying the limiting reagent.

2.   For each of the following reactions, write a balanced equation and then determine how many grams of product can be produced from 100 g of the first reactant. Assume you have as much of the second reactant as you need to use all of the first reactant.

a. Calcium oxide reacts with carbon dioxide to form calcium carbonate.

b. Iron plus oxygen react to form iron (III) oxide.

Cooperative Group Problems

1.   For each of the following reactions, write a balanced chemical equation, determine the limiting reagent, calculate how much product you can make, and calculate how much of which starting material will be left over when the reaction is complete.

a.   10.5 g of hydrogen reacts with 41.2 g of nitrogen to form ammonia.

b. 15.6 g of silicon reacts with 70.0 g of chlorine to form silicon tetrachloride.

2.   These problems all deal with percent yield, so they all use the same formula for their solution.

a. A chemist runs the reaction in Problem 1b to make silicon tetrachloride and isolates 85.3 g of product. What was the percent yield? Is that reasonable? What do you think might be wrong with this experiment?

b. Mary needs 56 g of gallium nitride, which can be made by sonochemical techniques with an 85% yield. How much gallium metal must she use to get the desired amount of product?

c. James tried to make 156 g of lithium oxide through a procedure that gave a 56% yield. How much lithium oxide did he make?

Practice Exam Questions
Unit 8

1. Consider the following equation. If 6.0 moles of $Al_{(s)}$ were used, how many moles of $MnO_{2(s)}$ would be needed to react completely?

$$3 MnO_{2(s)} + 4 Al_{(s)} \rightarrow 3 Mn_{(s)} + 2 Al_2O_{3(s)}.$$

    a. 4.0
    b. 5.0
    c. 4.5
    d. 5.5
    e. 4.3

2. Consider the following equation. If 3.86 g of $H_{2(g)}$ were used to react completely, how many grams of $C_6H_{12(g)}$ will be produced?

$$C_6H_{6(g)} + 3 H_{2(g)} \rightarrow C_6H_{12(g)}.$$

    a. 108 g
    b. 0.640 g
    c. 0.280 g
    d. 54.0 g
    e. 1950 g

3. Consider the following equation. If 35.5 g of $Cl_{2(g)}$ are used, how many moles of $KCl_{(s)}$ will be produced?

$$Cl_{2(g)} + 2 KI_{(aq)} \rightarrow I_{2(s)} + 2 KCl_{(aq)}.$$

    a. 0.500
    b. 1.00
    c. 1.50
    d. 2.00
    e. 2.50×10^3

4. Consider the following equation. If 40.0 g of $O_{2(g)}$ is used, how many grams of $C_2H_5OH_{(l)}$ will be needed to react completely?

$$C_2H_5OH_{(l)} + 3 O_{2(g)} \rightarrow 2 CO_{2(g)} + 3 H_2O_{(l)}.$$

    a. 173
    b. 9.28
    c. 2.00×10^4
    d. 19.2
    e. 9.10×10^{-3}

5. According to the following equation, 65.4 g of $H_2O_{(l)}$ was produced. How many moles of $HCl_{(aq)}$ were consumed during the reaction?

$CaCO_{3(s)}$ + 2 $HCl_{(aq)}$ → $CaCl_{2(aq)}$ + $H_2O_{(l)}$ + $CO_{2(g)}$.

    a. 7.27
    b. 3.61
    c. 1.80
    d. 5.90 x 10^2
    e. 2.40 x 10^3

6. In the following equation, 10.0 g of $NaHCO_{3(s)}$ and 15.0 g of $HCl_{(aq)}$ are used. How many grams of $CO_{2(g)}$ will be produced in this reaction?

$NaHCO_{3(s)}$ + $HCl_{(aq)}$ → $NaCl_{(aq)}$ + $H_2O_{(l)}$ + $CO_{2(g)}$.

    a. 18.1 g
    b. 7.86 g
    c. 12.2 g
    d. 5.24 g
    e. 6.23 g

7. Consider the following equation. If 2.3 g of $NH_{3(g)}$ are used and 4.5 g of $CO_{2(g)}$ are used, how many grams of $CN_2H_4O_{(s)}$ will be produced?

2 $NH_{3(g)}$ + $CO_{2(g)}$ → $CN_2H_4O_{(s)}$ + $H_2O_{(l)}$.

    a. 6.1 g
    b. 8.1 g
    c. 5.9 g
    d. 6.8 g
    e. 4.1 g

8. In the following reaction, 7.53 g of $Ca(OH)_{2(aq)}$ were used along with 10.5 g of $HBr_{(aq)}$. How many grams of $CaBr_{2(aq)}$ will be produced?

$Ca(OH)_{2(aq)}$ + 2 $HBr_{(aq)}$ → $CaBr_{2(aq)}$ + 2 $H_2O_{(l)}$.

    a. 19.0 g
    b. 9.30 g
    c. 11.8 g
    d. 20.3 g
    e. 13.0 g

9. A student carried out the reaction in Question 8 in the lab. He weighed the product and only 9.98 g of $CaBr_{2(aq)}$ was produced. Using your answer in Question 8 and your theoretical yield, calculate the percent yield.

    a. 7.61%
    b. 56.7%
    c. 77.0%
    d. 92.4%
    e. 3.41%

10. In the following reaction, 5.00 g of $UO_{2(s)}$ were used. In the lab, a technician obtained 3.21 g of UF_4 product. What is the percent yield of $UF_{4(aq)}$?

$$UO_{2(s)} + 4\,HF_{(aq)} \rightarrow UF_{4(aq)} + 2\,H_2O_{(l)}$$

    a. 45.6%
    b. 55.3%
    c. 75.3%
    d. 67.9%
    e. 32.5%

Unit 9:
Have You Seen My Electron?

If you are searching for enlightenment, it is often a good strategy to study the simplest systems possible. Removing complexity allows the most fundamental principles and behaviors to be observed. Neils Bohr studied the work of many other scientists, then he looked at data for the hydrogen atom. By studying this simple one-proton–one-electron system, he was able to deduce several shocking but irrefutable conclusions about the nature of electrons within atoms. Neils Bohr taught us that electrons in atoms have fixed energies. This quantization of energy was a totally revolutionary idea. Before then, people believed that any amount of energy could be added to a system. Now it was clear that electrons could only have certain amounts of energy. This information, combined with the knowledge that electrons behave as waves, allowed Erwin Schrodinger to write and solve a fascinating equation to describe electron behavior. The solutions of this equation are also equations, called wavefunctions, because they describe the wave behavior of electrons. My former instructor has described the remarkable correspondence between Schrodinger's solutions and the periodic table as, "The most beautiful thing in the world."

Fireworks

One of the most universally popular forms of entertainment is a brilliant display of fireworks. On the Fourth of July, students often ask what causes the colors of fireworks. When any element is heated to a high enough temperature, it will absorb heat energy and then release some of this energy in the form of light. Different elements release different energies of light; thus, they create different colors. Different colors of light are just light of different energies. An explosion containing lithium will emit pink light (low in energy), while an explosion containing copper will emit green light (higher in energy). After studying this unit, you should understand this process well.

Electromagnetic Radiation

Several different energy waves are different arbitrary classifications of one basic form of energy. X-rays, microwaves, radio waves, visible light rays, and others are all different types of electromagnetic radiation. The difference between them is that they carry different amounts of energy per photon. A photon is the carrier of electromagnetic radiation. In the television show, *Star*

Trek™, photon torpedoes are used to deliver a concentrated blast of electromagnetic radiation to specific targets.

Roy G. Biv

Roy G. Biv is a very common mnemonic that helps students remember the energetic ordering of the colors of light. White light can be separated into these component colors using a prism. In order of increasing energy per photon, the colors are red, orange, yellow, green, blue, indigo, and violet. Ultra-violet light is higher in energy than violet light; it is named because ultra means "even more." Infrared radiation is lower in energy than red light; it is named because infra means "below." These linguistic clues can help you understand the relative energy ordering of the electromagnetic spectrum. Another clue is the relative level of danger associated with each form of radiation. Ultraviolet radiation, X-rays, gamma rays, and cosmic rays are increasingly dangerous because they are the high-energy forms of radiation. Radio waves, microwaves, and infrared radiation can be used in communication devices, because in small doses, their energy is too low in energy to cause damage to our tissues.

Radio waves	Micro waves	Infrared	ROYGBIV	Ultra-violet	X-rays	Gamma rays

Low Energy   High Energy
Radiation   Radiation

Frequency and Wavelength

Any photon of light will be characterized by three distinct quantities: its energy, frequency, and wavelength. If you know any one of these quantities for a photon of electromagnetic energy, you can calculate the other two using the following two equations:

$$E = h\nu$$
$$\text{and}$$
$$c = \lambda \nu,$$

where c is the speed of light (3.0×10^8 m/s), h is Planck's constant (6.626×10^{-34} J·s), λ (*lambda*) is wavelength, and ν (*nu*) is frequency. By plugging the values of the constants c and h, you can solve these equations. High energy light must have a high frequency, but a short wavelength. This is a direct result of the preceding mathematical equations.

Direct and Indirect Proportions

The two preceding equations represent two very common types of equations. The first is a direct proportion, and the second is an indirect proportion. In a direct proportion, when the value of one variable increases, the value of the other must also increase to maintain the equality. $E=hv$ is one such proportion. If you increase the frequency of a photon, you will also increase its energy. The equation $c=\lambda v$, on the other hand, is an indirect proportion. If the wavelength of a photon increases, the frequency must decrease such that the speed of light remains equal to 3.0×10^8 m/s. If you do not remember direct and indirect proportions from your study of algebra, solving several numerical problems will help improve your understanding of these important relationships.

The Shocking Line Spectra

We have all seen that light can be separated into a beautiful rainbow by a prism. A device called a diffraction grating will also separate light in the same way. Years ago, physicists learned of a new way to produce light. They realized that they could place a small amount of pure gas into a glass tube and then send electricity through the gas. Thus, if the physicists used hydrogen gas, they could create a hydrogen lamp. A hydrogen lamp is just like the neon lights of Las Vegas, except it uses hydrogen instead of neon. When the light emitted from either of these sources was separated using a prism or diffraction grating, only a few distinct colors of light were seen in thin lines. Most of the rainbow-colored spectrum was missing. In fact, much of the invisible radiation (such as ultra-violet radiation and infrared radiation) was also missing.

Simply Beautiful

The spectrum of radiation produced by the hydrogen lamp contained lines with a pattern.

Figure 9.1 Line spectrum of hydrogen.

After years of interpretation, physicists were able to write one equation that summarized the energies of any of the lines produced by the hydrogen lamp. It is

$$\text{Equation 1:} \quad E = 2.18 \times 10^{-18}\left(\frac{1}{n_1^2} - \frac{1}{n_2^2}\right) \text{ J,}$$

where n_1 and n_2 are positive integers. This is truly an amazing equation. Because of its complexity, the obvious challenge was to explain *why* this equation described the energies of the individual hydrogen lamp lines so well.

Nobel Laureate Neils Bohr

Neils Bohr put together a series of assumptions that not only explained the line spectra of the hydrogen atom, but also gave us a first understanding of how electrons really behave in an atom. The true details of electron behavior may forever elude us, but after Bohr finished his work, we know more than we did. Bohr made three critical correct assumptions:

1. Electrons in an atom are at fixed energy levels (quantized).
2. Electrons can change from one energy level to another by either absorbing or emitting an exact amount of energy, which corresponds to the difference in energy between the two levels.
3. The line spectra are the result of light energy emitted as electrons undergo transitions from higher energy levels to lower ones, releasing their excess energy.

Thus, he derived the equation for an electron within the nth energy level of a hydrogen atom as

$$\text{Equation 2:} \quad E_n = -2.18 \times 10^{-18} \left(\frac{1}{n^2} \right) \text{ J.}$$

After this equation was derived, then people felt they understood the source of line spectra. A difference in energy between any two energy levels can be calculated using the energy of the final energy level minus the energy of the first energy level.

The Zero Point

We might wonder why the energy of an electron in the hydrogen atom is negative. The reference point for the energy of an electron is a free, isolated electron. When there are no forces acting on an electron, it is at zero energy. As it moves closer to the nucleus, it becomes more stable. And a more stable electron results in lower energy. Energies that are lower than the zero energy point of the free electron have negative energies. Removing an electron from an atom requires energy because of the attraction of oppositely charged particles. On the other hand, electrons that have been moved close to other electrons are high in energy. They can do work as they push away from the other electrons.

Bohr's Incorrect Assumptions

Although Neils Bohr was absolutely correct about the fixed energy of an electron within an atom, he made a few assumptions that were not correct. For example, he believed that each energy level could have one electron in it, and that the electron would orbit the nucleus at a fixed radius. This would lead to a circular electron orbit for each energy level, with the higher energy electrons being in larger orbits. Unfortunately, electron behavior is not so simple and predictable. The total energy of an electron in orbit is constant, but that energy is a combination of both potential and kinetic energy. An electron can move closer to the nucleus, where it is lower in potential energy if it gains speeds and, thus, rises in kinetic energy. The electron will not be confined to a fixed radius from the nucleus; it unpredictably moves closer to and farther from the nucleus. Also, because the electron behaves as a wave and not as a particle, most of us have an overly simplistic picture of an electron. If you continue to study chemistry, this simple particulate picture of the electron will be replaced with a more difficult one, but one that is ultimately more accurate.

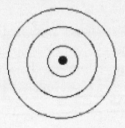

Figure 9.2 Bohr hydrogen orbital picture.

The fixed orbit assumption was one of Bohr's incorrect assumptions. He also incorrectly assumed that only one electron could have a particular energy level in the atom. As we will learn in the next unit, even ten electrons in an atom can all have the same energy. Sadly, the world is more complicated than Neils imagined.

Orbitals

The solutions to the Schrodinger equation provide a better picture of where electrons are located. If we graph the solutions of the Schrodinger equation in three dimensions, we will get shapes. Many people believe that the electron described by each solution to Schrodinger's equation spends most of its time in a region of space shaped like the graph. For the duration of this course, we will call these shapes "orbitals." We will think of them as places for electrons within an atom. Although many scientists will disagree with this simple picture, this interpretation led to great achievements in predicting the way chemicals react.

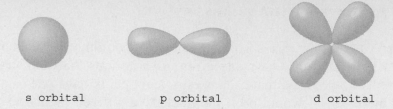

s orbital   p orbital   d orbital

Figure 9.3 Schrodinger hydrogen orbital shapes.

The shapes of orbitals are identified by letter designations. The Figure 9.3 lists the letter designations for the orbitals. The simplest one (a sphere) is called an s orbital. In order, the orbital shapes are s, p, d, f, and g. We will primarily be concerned with the s, p, d, and f orbitals.

Sublevels

There is a **lot** of new vocabulary in this section. It may take you a while to learn this vocabulary well enough to be able to understand your instructor.

A set of orbitals of the same shape is called a sublevel. The number of sublevels contained in an energy level depends upon the energy level itself. The first energy level has one sublevel, the s sublevel. The second energy level has two sublevels, the s and p sublevels. The third energy level has three sublevels, and so on.

Numbers of Orbitals in a Sublevel

Each s sublevel is composed of exactly one orbital. Each p sublevel is composed of exactly three orbitals. Each d sublevel is composed of exactly five orbitals. This is very important to remember. The reasons should become clear as you study this subject further.

Spin

Exactly two electrons can occupy each orbital. Although the electron does not really spin as was once believed, scientists still refer to the two electrons in an orbital as "spin up" and "spin down."

Addresses for Electrons

One useful way to think about the quantum numbers is that an energy level, energy sublevel, orbital, and spin tell us an address for an electron. Only one electron can have each address.

Ground State Electron Configurations

Ultimately, the reason we are concerned with the quantum numbers is that they will help us to understand the behavior of atoms and molecules. The electron's orbital will dictate the likelihood that the electron will react or engage in chemical bonds. We have already studied atomic composition, so we know how many electrons are in each atom or ion. Now we are ready to describe where these electrons are located. Under normal circumstances, electrons will occupy the lowest energy orbitals that are available. If all electrons are in the lowest energy orbitals, then they are said to be in "the ground state." The lowest energy orbital available in any atom is the 1s orbital. Like other orbitals, this orbital can hold two electrons. The next lowest energy orbital is the 2s orbital. After the 2s orbital is the set of three 2p orbitals. For the oxygen atom, the ground state electron configuration is $1s^2 2s^2 2p^4$. This means that there are 2 electrons in the 1s orbital, 2 electrons in the 2s orbital, and 4 electrons in the 2p orbitals; there are a total of 8 electrons.

Shorthand Notation for Core Electrons

If we had to assign the 83 electrons of Bi, just writing out the electron assignments would take us a long time. Since chemists are mainly concerned with the valence electrons (also called outer electrons), a shorthand method was developed. Any set of filled orbitals that correspond to the electron configuration of a noble gas are referred to as core electrons. These are generally un-reactive, and they are also called the inner electrons. If we want to write out the ground state electron configuration for Si, we could write $1s^2 2s^2 2p^6 3s^2 3p^2$ or we could write $[Ne]3s^2 3p^2$. The agreed-upon shorthand method is that [Ne] is the same as the ground state electron configuration of neon, $1s^2 2s^2 2p^6$. The same would be true for other noble gas electron configurations.

Using the Periodic Table

It would be virtually impossible for us to predict the detailed orbital energy ordering at this stage. Luckily, we do not have to, as the periodic table can be used as a map of the orbital energy ordering.

Two, Six, Ten, Fourteen

The periodic table was assembled before Schrodinger solved the equation intended to describe electrons. Chemical reactivity and the combining ratios of elements in compounds allowed Mendeleev and others to create the periodic table as we know it. Notice that the left section of the periodic table is 2 columns wide. The right section is 6 columns wide. The center section is 10 columns wide, and the separate section is 14 columns wide. It is more than a coincidence that the numbers of electrons that can fit in the s, p, d, and f orbitals are 2, 6, 10, and 14 respectively. The electrons control chemical behavior; the orbitals that the electrons occupy have a lot to do with how they react. The various sections of the periodic table are named the s, p, d, and f blocks.

The Map in Action

Examine a copy of the periodic table and compare it to the following diagram, which predicts how many electrons that an atom will have in its highest energy sublevel, as well as which sublevel is the highest one with electrons in it. As we follow the elements in order by atomic number, we move back and forth between the different blocks of the periodic table. We start in the s block for the first four elements; then we move into the p block for the next six. We return to the s block for two elements; then we go back to the p block for 6 more. Elements in the p block will have their outermost electrons in p orbitals. As we saw earlier, the electron configuration of oxygen is $1s^2 2s^2 2p^4$. We put the last 4 electrons of oxygen in the 2p orbitals and oxygen is in the fourth column of the p block of the table. The location of an element in the periodic table tells us how many electrons are in the outermost orbital and which type of orbital is the outermost orbital. Compare the following table to the periodic table to become more familiar with this concept.

Map of Predicted Electron Configurations, Outermost Electrons

$1s^1$																	$1s^1$
$2s^1$	$2s^2$											$2p^1$	$2p^2$	$2p^3$	$2p^4$	$2p^5$	$2p^6$
$3s^1$	$3s^2$											$3p^1$	$3p^2$	$3p^3$	$3p^4$	$3p^5$	$3p^6$
$4s^1$	$4s^2$	$3d^1$	$3d^2$	$3d^3$	$3d^4$	$3d^5$	$3d^6$	$3d^7$	$3d^8$	$3d^9$	$3d^{10}$	$4p^1$	$4p^2$	$4p^3$	$4p^4$	$4p^5$	$4p^6$
$5s^1$	$5s^2$	$4d^1$	$4d^2$	$4d^3$	$4d^4$	$4d^5$	$4d^6$	$4d^7$	$4d^8$	$4d^9$	$4d^{10}$	$5p^1$	$5p^2$	$5p^3$	$5p^4$	$5p^5$	$5p^6$
$6s^1$	$6s^2$	$5d^1$	$5d^2$	$5d^3$	$5d^4$	$5d^5$	$5d^6$	$5d^7$	$5d^8$	$5d^9$	$5d^{10}$	$6p^1$	$6p^2$	$6p^3$	$6p^4$	$6p^5$	$6p^6$

Figure 9.4 Predicted outermost electron configurations.

Box Form

In addition to writing electron configurations like $1s^2 2s^2 2p^4$, we can represent electron configurations in a more graphical way, as shown in Figure 9.5. With this method, each orbital is drawn as a box, and each electron is drawn as an arrow in the box. The boxes are drawn in a pattern so the energy ordering of the orbitals is clear.

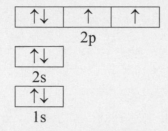

Figure 9.5 Graphical representation of oxygen electron configuration

Periodic Trends

Your book has a very good description of how the relative size and ionization energy of the elements can be predicted using the periodic table. Read the appropriate sections of the textbook prior to your group meeting.

Pre-workshop Problems

1.   The energies of photons (the carriers of electromagnetic radiation) are listed in the following questions.

a.   Complete the table by calculating each frequency, wavelength, and the region of the electromagnetic spectrum in which each photon is located.

Energy	Frequency	Wavelength	Region
6.0×10^{-19} J			
7.9×10^{-20} J			
4.4×10^{-25} J			

b.   As the energy of the photons decreased going down the table, what happened to the frequency and wavelength of the radiation? Use complete sentences to answer the question.

2. Using your textbook, write definitions for each of the following terms:

a. photon

b. wavelength

c. frequency

d. quantized

e. emission

f. electronic transition

3. In which column of which block of the periodic table would you find each of the following elements? It may help to use the periodic table that is found in this unit.

a. calcium

b. uranium

c. iodine

d. cesium

e. gold

f. lead

Cooperative Group Problems

1a.   According to Bohr's model of the H atom, what happens to the relative differences in energy between electron energy levels as you move to higher electron energy levels?

b.   Without doing any calculations, determine which of the following electronic transitions of the hydrogen atom will result in the emission of light of the highest energy? It may help to sketch a graph of relative energy levels based on your answer to Part a.

n=1 → n=4

n=2 → n=6

n=5 → n=2

n=2 → n=1

c.   Which of the transitions in Question 1b would result in the emission of light of longest wavelength?

2. Consider the following types of electromagnetic radiation:

yellow light,   blue light,   microwave,   infrared.

a. Which type has the highest frequency?

b. Which type has the longest wavelength?

c. Which type carries the most energy?

3a. How many sublevels exist for the fourth energy level? What are they?

b. For each sublevel, how many orbitals are in that sublevel?

c. Within an atom, how many total electrons can occupy the fourth energy level?

4.   Have each person in your group choose one of the following elements and write its ground state electron configuration on the blackboard; include both the line and box forms. When you are done, have each person explain how the problem is solved and whether or not Hund's Rule is illustrated in the example. Finally, check your textbook to make sure that everyone has the correct electron configuration. If any electron configurations are incorrect, determine the correct ground state configurations.

a. argon

b. titanium

c. phosphorus

d. magnesium

e. iron

f. aluminum

g. copper

h. chromium

i. silicon

j. chlorine

Practice Exam Questions
Unit 9

1. Which of the following terms is not a characteristic of electromagnetic waves?

> a. wavelength
> b. particles
> c. frequency
> d. speed
> e. all of the above are characteristics

2. The distance between two consecutive peaks in a wave is the

> a. amplitude
> b. frequency
> c. photon
> d. wavelength

3. The number of cycles per second is called the

> a. frequency
> b. amplitude
> c. speed
> d. wavelength

4. When an atom is said to be in the excited state, it has

> a. an unlimited amount of energy
> b. very little energy
> c. excess energy
> d. no energy
> e. nothing to do with energy

5. When an electron moves from a higher energy orbital to a lower one,

> a. energy is absorbed.
> b. light is emitted.
> c. the electron becomes less stable.
> d. the atom increases in size.
> e. the pressure increases

6. What is the frequency of a photon with a wavelength of 3.02×10^{-2} cm?

    a. 9.06×10^{6} s^{-1}
    b. 9.06×10^{4} s^{-1}
    c. 9.93×10^{9} s^{-1}
    d. 9.93×10^{11} s^{-1}
    e. 9.06×10^{8} s^{-1}

7. What is the wavelength of a photon with a frequency of 6.53×10^{14} s^{-1}?

    a. 1.96×10^{23} m
    b. 1.96×10^{23} nm
    c. 1.96×10^{14} nm
    d. 4.59×10^{-7} nm
    e. 459 nm

8. What is the energy of a photon with a wavelength of 6.7×10^{-6} m?

    a. 4.4×10^{-39} J
    b. 4.4×10^{-39} kJ
    c. 4.4×10^{-39} J
    d. 4.5×10^{13} J
    e. 3.0×10^{-20} J

9. What is the wavelength of a photon that carries 4.67×10^{-15} J of energy?

    a. 4.26×10^{-11} m
    b. 7.05×10^{18} m
    c. 3.09×10^{-48} m
    d. 4.26×10^{-11} nm
    e. 7.05×10^{18} nm

10. A transition between which electronic energy levels in the hydrogen atom result in an emission of the highest energy?

    a. n=1 → n=3
    b. n=3 → n=2
    c. n=1 → n=7
    d. n=2 → n=1
    e. n=7 → n=5

11. A transition between which electronic energy levels in the hydrogen atom result in an absorption of the longest wavelength?

     a. n=1 → n=3
     b. n=3 → n=2
     c. n=1 → n=7
     d. n=2 → n=1
     e. n=7 → n=5

12. The probability map for the hydrogen electron is called

     a. a principal energy level
     b. an orbital
     c. a sublevel
     d. the periodic table
     e. a mechanical model

13. Which element has the following ground state electron configuration?

$1s^2 2s^2 2p^6 3s^2 3p^2$

     a. Al
     b. Si
     c. P
     d. C
     e. Ge

14. How many 3d electrons are found in copper?

     a. 1
     b. 2
     c. 8
     d. 9
     e. 10

15. How many valence electrons does selenium have?

     a. 2
     b. 3
     c. 4
     d. 5
     e. 6

16. How many 5p electrons does tellurium have?

    a. 0
    b. 4
    c. 5
    d. 6
    e. 8

17. What is the electron configuration for strontium?

    a. $1s^22s^22p^63s^23p^64s^23d^{10}4p^2$
    b. $1s^22s^22p^63s^23p^64s^2$
    c. $[Ar]4s^2$
    d. $[Ar]4s^23d^{10}4p^4$
    e. $[Kr]5s$

18. Which element has the following ground state electron configuration?
$1s^22s^22p^63s^23p^64s^23d^{10}4p^65s^24d^{10}5p^4$

    a. iron
    b. polonium
    c. tellurium
    d. xenon
    e. selenium

19. How many valence electrons does bromine have?

    a. 4
    b. 5
    c. 6
    d. 7
    e. 8

20. The electrons in the highest energy level of a ground state atom are the
    _____ electrons.

    a. core
    b. valence
    c. inner
    d. excited
    e. spherical

21. How many electrons can occupy the set of d orbitals in the fourth energy level of an atom?

    a. 0
    b. 2
    c. 5
    d. 10
    e. 32

Unit 10: Bonds Away

One of the biggest challenges for chemists was to figure out what holds chemical substances together. Physicists have compiled a list of the forces that generally cause attractions between two objects, and these forces are very useful for understanding chemical systems. They are electrostatic (between charged objects), magnetic (between objects with magnetic fields), gravitational (between objects with mass), and nuclear (between subatomic particles). Ultimately, all chemical bonds are electrostatic in nature, but how we visualize them is very different depending on the species involved. During this unit, we will investigate the two most important types of compounds, ionic and covalent, as well as the bonding in these types of compounds.

Salts and Ionic Bonds

If we combine a metal and a non-metal, we get a salt. Many salts will dissolve in water to create ionic species in solution. In this situation, it is clear that the metal has become positively charged and the non-metal has become negatively charged. Consider the reaction between sodium and chlorine to form sodium chloride:

$$2 \, Na_{(s)} \; + \; Cl_{2(g)} \; \rightarrow \; 2 \, NaCl_{(s)}.$$

As we discussed, we consider this an electron transfer reaction. Negatively charged electrons move from the sodium to the chlorine, making positively-charged sodium(1+) ions and negatively-charged chloride(-1) ions. Figure 10.1 illustrates the packing arrangement in an ionic compound like sodium chloride, which has a 1:1 ratio of cation to anion.

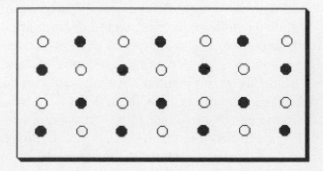

Figure 10.1 Representation of a sodium chloride lattice.

The sodium chloride salt dissolves in water, which will liberate the ions, as shown in the following equation. This dissociation into ions is evidence that the solid is composed of ions:

$$NaCl_{(s)} \xrightarrow{H_2O} Na^+_{(aq)} + Cl^-_{(aq)}.$$

If you evaporate the water, the attraction between the sodium cations and the chloride ions allows the solid to reform.

Lattice Energy

The attraction between oppositely charged ions is often called an ionic bond. The energy associated with this bonding is called the lattice energy. The amount of energy required to separate a salt crystal into its component ions in the gas phase is called the lattice energy. If the ions in a salt have higher charges, then they are more difficult to separate. MgO, which is composed of +2 and −2 charged ions, will have a higher lattice energy than NaCl, which is composed of +1 and −1 charged ions. To a lesser extent, the sizes of the ions will also affect the lattice energy of ionic compounds. Smaller ions cause the positive and negative charges to be closer to each other; thus, they are more difficult to separate.

Covalent Bonds

It is clear and fairly easy to understand why sodium ions are attracted to chloride ions. It is not as clear why hydrogen atoms would be attracted to each other and helium atoms would not be attracted to each other.   Two hydrogen atoms will form H_2, which is the only stable molecule for hydrogen, but oxygen can form either O_2 or O_3. Why should they be different? G. N. Lewis gave us an explanation to help explain both covalent and ionic substances. The basic idea is that every atom is stable once it has 8 valence electrons associated with it. There are two ways that an atom can get these electrons. The first way is to transfer electrons to or from another atom. The second way is to share electrons with another atom. Lewis was one of the first chemists to define a bond as a shared pair of electrons. When a bond results from the sharing of electrons, it is called covalent. Even a covalent bond is essentially due to charge attractions. Placing a pair of electrons between two nuclei will cause those nuclei to move closer to each other as they move toward the bonding electrons. Figure 10.2 uses the hydrogen molecule to depict the electrostatic attractions between the nuclei and electrons.

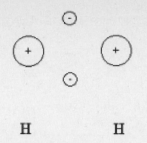

H   H

Figure 10.2 A hydrogen molecule showing the electrostatic interaction

Electronegativity and Bond Polarity

When two atoms are bonded together, one might pull electrons to itself with more force than the other. Linus Pauling invented electronegativity so that we could predict the type of bond that will form between two atoms. Electronegativity is the ability of an element to attract bonding electrons to it. If one atom pulls much harder than the other, the electrons get transferred to that atom. For example, chlorine has a very high electronegativity (3.0) and sodium has a very low electronegativity (0.9), so chlorine becomes negatively charged when those elements combine. If two hydrogen atoms form H_2, they both exert the same attractive force for the electrons (the electronegativity of H is 2.1), so the electrons between them are shared equally. If a hydrogen atom was bonded to a chlorine atom, then there is a medium-sized difference in the electronegativities between them. This difference is not strong enough to cause a full transfer of electrons from H to Cl, but it is strong enough to cause a partial transfer. The slight motion of electrons toward the chlorine atom puts a slight negative charge on the chlorine and leaves a slight positive charge on the hydrogen. This causes the bond to be polar, which means that one side of the bond has a slight negative charge and the other has a slight positive charge. Notice that electronegativities are lower for elements that are to the left and bottom of the periodic table. Even if we do not know exact values, we can use this trend to predict relative polarity of bonds.

F-F	O-F	N-F	C-F	B-F	Li-F
Non-polar	slightly polar	polar	very polar	very polar	ionic

Electronegativity of Main Group Elements

Li 1.0	Be 1.5	B 2.0	C 2.5	N 3.0	O 3.5	F 4.0
Na 0.9	Mg 1.2	Al 1.5	Si 1.8	P 2.1	S 2.5	Cl 3.0
K 0.8	Ca 1.0	Ga 1.6	Ge 1.8	As 2.0	Se 2.4	Br 2.8
Rb 0.8	Sr 1.0	In 1.7	Sn 1.8	Sb 1.9	Te 2.1	I 2.5
Cs 0.7	Ba 0.9	Tl 1.8	Pb 1.9	Bi 1.9	Po 2.0	At 2.2

Figure 10.3 Electronegativities of main group elements.

Many people find it useful to remember the relative ordering of the electronegativities of the common elements for the prediction of bond polarity:

$$F > O > N, Cl > Br > C, S > P, H.$$

Picture of Polarity

We are already familiar with one example of polarity. A bar magnet has magnetic polarity, which makes it behave similar to something with electric polarity. A pair of bar magnets will stick to each other because they can align so that there are attractions between the positive pole of one magnet and the negative pole of the other. Figure 10.4 illustrates this concept.

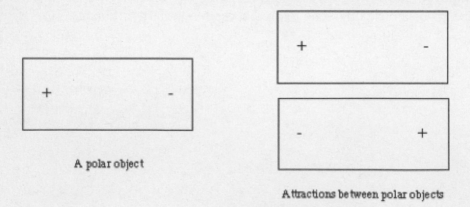

A polar object

Attractions between polar objects

Figure 10.4 Illustrations of polar objects and their interactions.

The Octet and Duet Rules

Valence electrons are the outermost electrons in an atom or molecule. Since they are on the outside of the atom, they are the electrons that can be involved in bonds with other atoms. The correct number of valence electrons required for each element to be stable in compounds depends on the row of the periodic table in which it is located. Hydrogen and helium only need two valence electrons. This is sometimes referred to as the duet rule. Elements in the second row (B, C, N, O, F, and Ne) will have eight valence electrons in most of their stable compounds and cannot have more than eight valence electrons. This tendency is referred to as the octet rule. Elements in the third or higher will also tend to have eight electrons, but this number can sometimes be greater than 8. If an atom in the third row or higher has greater than eight electrons, it is said to have an expanded octet or to have violated the octet rule.

Predicting Formulas

It was already known that the formulas for a series of compounds were HF, H_2O, H_3N, and H_4C. The pattern was recognized, but there was no explanation for why the formulas should be this way. Atoms in the noble gas group already have the stable number of valence electrons, so they neither transfer nor share electrons. Fluorine is one short of eight valence electrons, so it will either accept one electron or form one bond. Oxygen has two electrons fewer than eight valence electrons, so it will typically share twice to get two more electrons and thus form two bonds. Nitrogen will form three bonds and carbon will form four bonds. Noble gases already have eight valence electrons, so they are stable by themselves and do not normally form compounds. Electrons in Lewis Structures are written as dots, as shown in Figure 10.5.

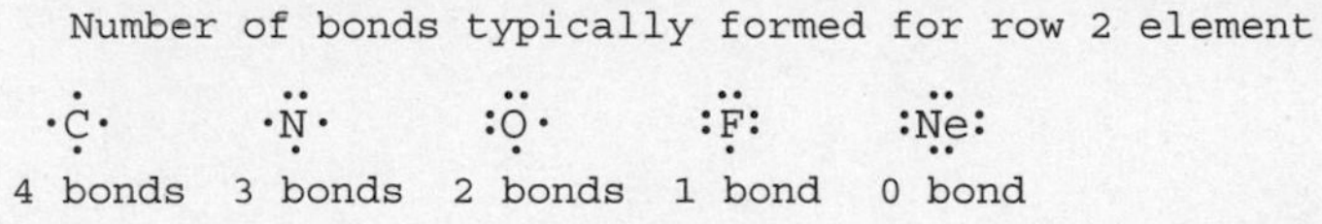

<pre>
       Number of bonds typically formed for row 2 element

      ·Ċ·   ·N̈·   :Ö·   :F̈:   :N̈e:
    4 bonds   3 bonds   2 bonds   1 bond   0 bond
</pre>

Figure 10.5 Lewis dot structures for atoms.

Drawing Lewis Structures

There are two popular methods for drawing Lewis Structures: the intuitive approach and an algorithmic approach. In general, the algorithmic approach is much easier for beginning students, so that is the method that we will use here. For any simple atom, molecule or ion, following the five-step procedure will generally lead to a good Lewis Structure.

1. Add the number of valence electrons in the molecule or ion.

2. Connect the atoms in a reasonable way to single bonds. Typically, there is a unique central atom to which the other atoms are attached.

3. Fill the octets on the non-central atoms by adding pairs of electrons to each one. Note that hydrogen is a special case because it is stable with two valence electrons. When it is singly-bonded to another atom, it has a filled duet of electrons.

4. Put unused electrons on the central atom.

5. If the central atom does not have an octet, use lone pairs of electrons on the non-central atoms to form multiple bonds to the central atom until the central atom has an octet of electrons.

You will get substantial practice drawing Lewis Structures when you complete the problems in this unit.

Pre-workshop Problems

1.   Draw a Lewis Dot diagram for each of the following atoms or ions. Which ones obey the octet rule?

a. Na

b. Na^+

c. P^-

d. O^{2-}

e. I^-

f. F

2.   Predict how many bonds each of the following atoms will typically form in stable complexes. Draw a Lewis dot diagram for each.

a. P

b. Cl

c. H

d. Ar

3. Predict formulas for compounds that would form between the following pairs of atoms. Draw Lewis Structures for each molecule you predict.

a. S and H

b. P and Cl

c. N and F

d. I and H

4. Each of the following diatomic molecules is stable. Draw a structure for each that obeys the octet rule and has the proper number of electrons. When you draw a stable Lewis Structure for each molecule, how many bonds are formed between the two atoms?

a. O_2

b. F_2

c. N_2

Cooperative Group Problems

1. Without looking up values for the electronegativity of each element, decide whether the electronegativity difference for each pair of elements is large or small. Explain your reasoning.

a. H and C

b. Si and F

c. Rb and Cl

d. O and F

2. For each of the following bonds, predict whether it will be a non-polar covalent bond, a polar bond, or an ionic bond?

a. H-F

b. S-F

c. Mg-Cl

d. I-I

3.   Draw a Lewis Structure for each molecule or ion. How many lone pairs of electrons are on each central atom?

a. water

b. hydrogen fluoride

c. phosphine (PH_3)

d. nitrate

e. oxygen difluoride

f. sulfate

g. sulfur trioxide

h. formaldehyde (CH$_2$O with C as the central atom)

i. sulfur tetrafluoride

j. sulfur hexafluoride

k. iodine trichloride

l. xenon difluoride

Practice Exam Questions
Unit 10

1. A polar covalent bond is the result of

   a. unequal sharing of electrons between atoms
   b. equal sharing of electrons between atoms
   c. the transfer of electrons from one atom to another
   d. magnetic fields
   e. none of the above

2. The periodic table trends for electronegativity are

   a. decreasing down, decreasing across
   b. increasing down, increasing across
   c. decreasing down, increasing across
   d. increasing down, decreasing across
   e. there is no pattern

3. Which bond is the most ionic?

   a. H-Cl
   b. Mg-O
   c. K-O
   d. F-F
   e. Si-F

4. Which bond is the least polar?

   a. H-Cl
   b. P-S
   c. S-F
   d. C-O
   e. Se-F

5. The polarity of a bond ____ as the difference in electronegativity between atoms ___.

   a. increases; increases
   b. stays the same; decreases
   c. increases; decreases
   d. polarity of a bond does not relate to electronegativity differences

6. How many valence electrons are in the nitrate ion?

    a. 1   d. 23
    b. 5   e. 24
    c. 22

7. This Lewis structure could be any element in what group?

$$:\overset{.}{X}:$$

    a. 4A   d. 7A
    b. 5A   e. 8A
    c. 6A

8. Which element is most likely to have a lone pair of electrons and bonds to three chlorine atoms?

    a. B   d. O
    b. C   e. F
    c. N

9. What is the N-N bond order in N_2H_2?

    a. 0   d. 3
    b. 1   e. 4
    c. 2

10. How many lone pairs of electrons are on the central S atom in SO_2?

    a. 0   d. 3
    b. 1   e. 4
    c. 2

11. Which has a triple bond?

    a. fluorine   d. nitrogen
    b. oxygen   e. ammonia
    c. sulfur dioxide

12. How many bonds is a carbon atom likely to form in a stable compound?

    a. 0   d. 3
    b. 1   e. 4
    c. 2

Unit 11: 2-D or not 2-D

Lewis structures provide considerable information about the bonding in molecules or polyatomic ions. The real, three-dimensional structures of molecules are important if we are to understand how they behave in our three-dimensional world. The interaction of drugs with their targets in the body is one example of how the three-dimensional structure of the molecules is of prime importance. Pharmaceutical companies hire teams of chemists to design molecules with a particular architecture for use as drugs. In order to understand modern medicine or drug design, it is essential that we understand molecular structure in three dimensions.

VSEPR

It is critical to understand that the behavior of electrons greatly influences the behavior and structure of chemicals. Electrons will always repel each other. Thus, the electrons in bonds on the central atom and lone pairs of electrons on the central atom will repel each other. This will cause them to move as far away from each other as possible. The theory based on this simple idea has a complex name: Valence Shell Electron Pair Repulsion. This theory builds on Lewis Structures. In order to apply VSEPR and determine the shape of a molecule, you must first draw an accurate Lewis Structure for that molecule.

The Perfect Shapes

If objects attached to a central point distribute themselves perfectly around a central point, the result is a collection of very symmetrical shapes known as the Platonic solids. We will need to be familiar with the five shapes pictured in Figure 11.1: the line, the trigonal plane, the tetrahedron, the trigonal bipyramid, and the octahedron. The table shows the number of things attached to a central point, the name of the electronic geometry, the bond angles in the structure, and a picture of the shape. Throughout this unit, we will refer to these shapes as the electronic geometries because they show the shape of a molecule when we include lone pairs in the picture. Many textbooks will use different terminology to refer to the electronic geometry. The original term is electron pair geometry. Additional terms are electron domain geometry or electron group geometry. Many people disfavor the term electron pair geometry because an atom only counts as one thing whether it is singly bonded or doubly bonded. It is important to note that there are only 5 electronic geometries. If we are ever asked to show the electronic geometry around a central atom, we have only five options.

Number of non-central things on the central atom (lone pairs of e⁻ or atoms)	Electronic geometry	Bond angles	Illustration
2	linear	180°	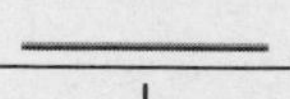
3	trigonal planar	120°	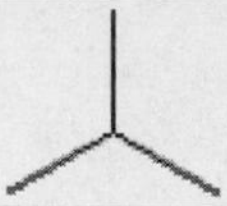
4	tetrahedral	109.5°	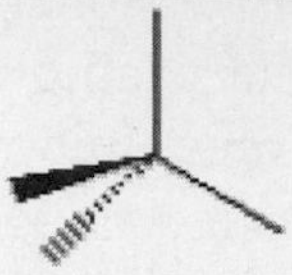
5	trigonal bipyramidal	90°, 120°, and 180°	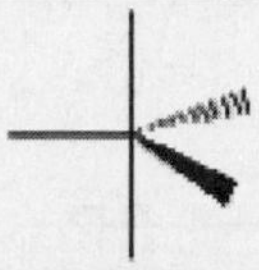
6	octahedral	90° and 180°	

Figure 11.1 The five electronic geometries.

Molecular Shapes

When the structure of a molecule is determined, the locations of the lone pairs of electrons are not directly determined. Structure determinations decide the positions of the core electrons of atoms, but not the valence electrons. We know that lone pairs are present on a central atom when the non-central atoms are closer to each other than they would be if only atoms were attached to the central atom. If there were only atoms on the central atom, the non-central atoms would be free to spread and become farther apart from each other. If lone pairs of electrons are present on the central atom, they require approximately the same amount of space as the atoms that are bonded to the central atom. When we describe the molecular shape or geometry of a molecule, we will describe the shape formed by only the atoms. The lone pairs on the central atom are important in determining the shapes of the molecules, but they are not visible in the structure. All of the molecular shapes are derived from the electronic geometries. Frequently, you will be asked to give both the electronic geometry and the shape for molecules or polyatomic ions. Learning the distinction between the two terms is important.

General Formulas

There are only a small number of possible shapes that can be formed from the parent electronic geometries. Using an abstract system, we can list all of them using the general formula AB_nE_m, where A is the central atom, B is a non-central atom, and E is a lone pair of electrons on the central atom. The electronic geometry will be determined by the sum $B + E$, the number of non-central atoms plus the number of lone pairs attached to the central atom. We will assign a name to each shape, some of which are different from the electronic geometries we already named. These shapes are listed in Figure 11.2.

General formula	B + E (atoms plus lone pairs on central atom)	Electronic geometry	Shape
AB_2E_0	2	linear	linear
AB_3E_0	3	trigonal planar	trigonal planar
AB_2E_1	3	trigonal planar	bent (angular)
AB_4E_0	4	tetrahedral	tetrahedral
AB_3E_1	4	tetrahedral	trigonal pyramidal
AB_2E_2	4	tetrahedral	bent (angular)
AB_5E_0	5	trigonal bipyramidal	trigonal bipyramidal
AB_4E_1	5	trigonal bipyramidal	sec-saw
AB_3E_2	5	trigonal bipyramidal	T-shaped
AB_2E_3	5	trigonal bipyramidal	linear
AB_6E_0	6	octahedral	octahedral
AB_5E_1	6	octahedral	square pyramidal
AB_4E_2	6	octahedral	square planar

Figure 11.2 Possible molecular shapes.

Predicting Shapes: One Example

Sulfur dioxide and carbon dioxide have similar formulas, but their structures are quite different. Carbon dioxide is linear, but sulfur dioxide is bent. The Lewis diagram for sulfur dioxide has one lone pair of electrons on the central sulfur atom, a single bond between one of the oxygen atoms and the sulfur, and a double bond between the other oxygen atom and the sulfur. The presence of the lone pair on sulfur pushes the two oxygen atoms closer to each other. The lack of a lone pair on the carbon allows the two oxygen atoms to move farther away from each other. Figure 11.3 illustrates this concept.

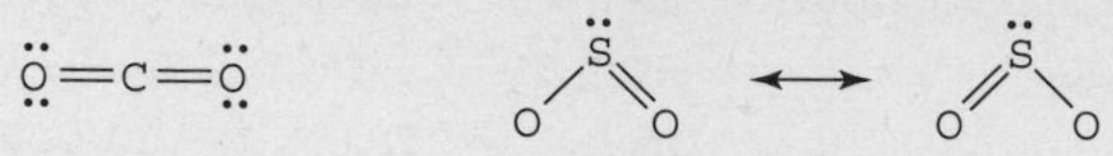

Figure 11.3 Lewis dot structures for carbon dioxide and sulfur dioxide.

Because the central sulfur atom contains a total of three things, the electronic geometry around that atom is trigonal planar. To visualize the molecular shape, we look at the geometry of just the atoms without the lone pairs. In this case, the geometry is bent with an ideal angle of 120 degrees from oxygen to sulfur to oxygen.

Pre-workshop Problems

1.   Draw a Lewis structure for each of the following atoms or ions. This will save you time when you need to deduce their structures later.

a. dihydrogen sulfide

b. nitrite

c. phosphorus trichloride

d. boron tribromide

e. carbon disulfide

f. ammonia

g. sulfur tetrafluoride

h. silicon tetraiodide

i. iodine trifluoride

Cooperative Group Problems

1.   At least two of the molecular shapes are called bent. Are they identical? If not, what makes them different?

2. In the "general shapes" portion of this unit, Figure 11.1 illustrates the molecular shapes that can be derived from the five electronic geometries. Draw each shape in perspective so the three-dimensional geometry is clear to an observer. Use wedges and dashed lines to represent the three-dimensional structures. Indicate bond angles between adjacent bonded atoms.

3. For each of the following molecules, list its electronic and molecular geometry. If you have a molecular model set available, build a model of the molecule.

a. dihydrogen sulfide

b. nitrite

c. phosphorus trichloride

d. boron tribromide

e. carbon disulfide

f. ammonia

g. sulfur tetrafluoride

h. silicon tetraiodide

i. iodine trifluoride

4. Consider the series of compounds BBr_3, NBr_3, and IBr_3. All have three bromine atoms attached to a central atom, but they have very different structures.

a. Draw the Lewis structure for each of these molecules.

b. What causes the molecular geometries of the three molecules to be different?

c. In general, when is the molecular geometry of a molecule identical to its electronic geometry?

Practice Exam Questions
Unit 11

1. The ideal bond angle in a tetrahedron is

    a. 90°
    b. 109.5°
    c. 120°
    d. 180°
    e. 270°

2. If the central atom has two atoms and one lone pair of electrons attached to it, what is the electron-pair geometry about the central atom?

    a. linear
    b. bent with a bond angle of 109°
    c. bent with a bond angle of 120°
    d. trigonal planar
    e. tetrahedral

3. If a central atom has two atoms and two lone pairs of electrons attached to it, what is the shape of the molecule?

    a. linear
    b. bent with a bond angle of 109°
    c. bent with a bond angle of 120°
    d. trigonal planar
    e. tetrahedral

4. The name that best describes the arrangement of electron pairs around the central As in AsH_3 is

    a. linear
    b. bent
    c. trigonal pyramidal
    d. trigonal planar
    e. tetrahedral

5. The molecular structure of boron trifluoride is

    a. linear
    b. bent
    c. trigonal pyramidal
    d. trigonal planar
    e. tetrahedral

6. The molecular geometry of sulfite is

        a. trigonal planar
        b. trigonal pyramidal
        c. tetrahedral
        d. square pyramidal
        e. T-shaped

7. The electron group geometry of sulfite is

        a. trigonal planar
        b. trigonal pyramidal
        c. tetrahedral
        d. square pyramidal
        e. T-shaped

8. What are the bond angles in carbonate?

        a. 90°
        b. 109.5°
        c. 120°
        d. 180°
        e. 360°

9. What is the H-C-C bond angle in C_2H_2?

        a. 90°
        b. 109.5°
        c. 120°
        d. 180°
        e. 360°

Unit 12: The Light Stuff

Without gases, no one would be alive right now. If we did not respire, we would soon expire. Oxygen is a gas that is essential for human life. However, despite the essential nature of oxygen, not all gases are good for us. Many pollutants are gases that threaten to make our planet uninhabitable. Nitrogen and sulfur oxides cause acid rain, and ozone threatens to make the air around most cities too toxic to support life. Because our lungs provide the quickest route to our bloodstream, toxic gases are particularly hazardous. Nitrogen oxides, sulfur oxides, and ozone burn lung tissue, causing terrible damage and pain to those who are exposed to them. On a more positive note, drug manufacturers are now trying to develop inhalable drug delivery systems that take advantage of the rapid uptake of gases into our bloodstream. Understanding the general properties of gases will help you to handle them in the laboratory and to understand how they affect you in your daily life.

Fly and be Free

Gas particles have enough energy to literally fly like Superman. They travel uninhibited in a straight-line path until they hit a barrier. If there is open space, gases will eventually travel everywhere within that space as they carom off each other and the walls of their container. Unlike Superman, gas particles travel without a purpose.

Direct and Indirect Proportions

When we studied the properties of electromagnetic radiation, we briefly encountered simple relationships between two variables: direct and indirect proportions (see the second page of Unit 9). As we study gases, we will use these proportions frequently. Algebra is critical in this unit. Many of the gas laws simply state the relationship between two quantities that we measure when we want to describe a gas. Gases are described by the amount of gas in moles (n), the pressure the gas is under (P), the temperature of the gas (T), and the volume it occupies (V). Boyle's Law states that there is an indirect proportion between the pressure and volume of a gas if it is at a constant temperature and the amount of gas remains constant. Boyle's Law can be described in multiple modes. There is a verbal description of the law, an algebraic description, and a graphical description. There is also a proportion that follows from the algebraic description of a gas law. Charles's Law can also be described in these multiple modes. Charles's Law states that the volume of a gas is directly proportional to its temperature when the pressure and amount of the gas are held constant.

Charles's Law

Verbal: Under conditions of constant pressure and the number of moles of a gas, the temperature is directly proportional to the volume of the gas.

Algebraic: $T = kV$ (constant P and n).

Proportion: $\dfrac{T_1}{T_2} = \dfrac{V_1}{V_2}$. (This form is often very convenient to use.)

Graphical Representation

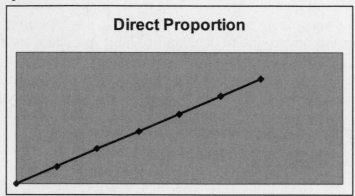

Figure 12.1 Graph of a direct proportion.

Restrictions

Most of the gas laws are only correct under very specific and controlled circumstances. In order to study the relationship between the pressure of a gas and its volume, the temperature of a gas and the amount of gas must be held constant. If not, multiple factors would affect the volume of the gas, not just its pressure. It is very important for you to pay close attention as to when a particular gas law is valid.

Gas Law	Variables	Constants
Charles's	*T, V*	n, P
Boyle's	*P, V*	n, T
Avogadro's	*n, V*	P, T

Pressure

The gases of the atmosphere are constantly pressing down on the Earth due to their mass. The resultant force is equal to 14.7 pounds per square inch. This pressure is commonly measured by letting the weight of the atmosphere push on a surface of mercury, which causes the mercury to flow up into a tube. If nothing pushes back, the mercury is lifted to a height of approximately 760 mm. The unit of a standard atmosphere has been accepted to be equal to

760.0 mm of mercury, which is the lifting power of standard atmospheric pressure. In honor of a famous scientist, Torricelli, who learned a great deal about gas pressure, the unit of 1 mm of Hg is named a Torr.

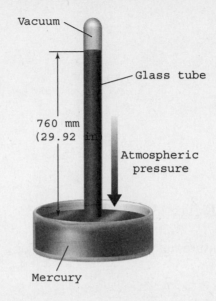

Figure 12.2 A mercury barometer.

Atmospheric pressure causes a column of Hg to be lifted.

760 mm Hg =760 Torr = 1 atm = 14.7 lb/in^2.

Measurement of Pressure

The measurement of pressure always consists of a gas sample pushing against something. In a U-tube manometer, one gas pushes down on one side of a column of mercury while another pushes down on the other side. The difference between the heights of the two sides of the mercury corresponds to the difference in pressure between the two gases.

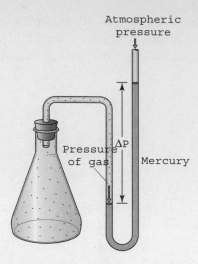

Figure 12.3 A U-tube manometer.

Gas pressure pushes against atmospheric pressure.

Using Gas Laws

Gas laws are frequently used to predict a final measurement of a gas after it has undergone a change of conditions. If 25 L of a gas is at a temperature of 250 K and it is heated to a temperature of 300 K, we can predict a final volume for the gas using Charles's Law. The initial conditions of the gas are T_1 and V_1. The final temperature is T_2. We can use this proportion to write

$$\frac{T_1}{T_2} = \frac{V_1}{V_2} \qquad \frac{250K}{300K} = \frac{25L}{V_2}.$$

Then solving for the final volume, V_2, is just a matter of algebra: V_2 is 30 L. It is extremely important to use temperatures in Kelvin when doing calculations involving gases.

The Ideal Gas Law

Each of the gas laws teaches something about gas behavior. When these laws are put together algebraically, the result is an equation called the ideal gas law, or

$$PV = nRT,$$

where P is the pressure of the gas, V is the volume of the gas, n is the number of moles of the gas, T is the temperature of the gas, and R is the proportionality constant that holds the equation true. R is equal to 0.082 L atm/(mole K). This equation and the constant may look formidable at first, but you must realize that it is simply an equation of four variables and one constant. If you know the value of three variables, you can always solve for the fourth. Chemists commonly calculate the number of moles of a gas by measuring the pressure, volume, and temperature of the gas.

Changing Conditions

$PV=nRT$ is a very useful equation for understanding a gas at a particular time. Many times, we will want to know how changing conditions will alter the properties of a gas. At the beginning of a study, we can measure the initial conditions of a gas and write $P_iV_i=nRT_i$, where the subscript i stands for initial conditions. After conditions have changed, $P_fV_f=nRT_f$ will be true, where the subscript f stands for final conditions of the gas. If both of those equalities are true, we can write an equation called the combined gas law:

$$\frac{P_iV_i}{P_fV_f} = \frac{n_iRT_i}{n_fRT_f} \text{ and then } \frac{P_iV_i}{P_fV_f} = \frac{T_i}{T_f}.$$

Note that the constant R and the number of moles of gas both cancelled during the derivation of this equation. This assumes that our container is closed and does not allow gases to leave or enter the container, so the values of n before and after the change are equal ($n_i = n_f$). Often, this equation will be rearranged so that all of the initial conditions are on the left side of the equation and the final conditions are on the right side of the equation:

$$\frac{P_iV_i}{T_i} = \frac{P_fV_f}{T_f}.$$

Standard Conditions

Chemists and physicists have defined standard conditions for a gas (STP) to be 1 atmosphere of pressure and 0°C. Zero degrees Celsius was an easy temperature to maintain even before refrigeration because ice water will stay at that temperature.

Kelvin is for Gases

Whenever you want to perform a calculation on a gas, you must convert the temperature to Kelvin. Although -100°C may seem freezing cold, there is plenty of energy available to a gas at this temperature. At lower temperatures, gases move more slowly than at room temperature, but they still have enough energy to fly. Only at zero degrees on the Kelvin scale is there no energy available for molecular motion.

Understanding

The gas laws tell us what happens to gases, but they do not tell us why gases behave as they do. However, there is a particulate picture of gases that does help explain their behavior. The Kinetic Molecular Theory consists of five postulates that explain the behavior of gases and describe characteristics of ideal gases. Different texts phrase these postulates differently, group them in different ways, or number them differently, but the general meaning of these postulates is the same.

1. Gases are composed of many small particles (atoms or molecules) in continuous random motion. Collisions between the molecules of a gas and the walls of its container cause the pressure exerted by the gas.

2. The sum of the sizes of the gas particles is very small compared to the distances between the gas particles. In other words, the size of the container is much larger than the sum of the sizes of the gas molecules. The molecules in liquids and solids essentially touch each other and do not have any space between them.

3. The particles neither attract nor repel each other. In more technical terms, inter-particle forces (attractions and repulsions) are negligible for gases.

4. The average kinetic energy (and the speed) of the gas is directly proportional to the temperature of the gas in Kelvin.

5. Collisions between gas molecules are elastic. This means that energy is transferred between gas molecules, but the total energy does not change during collisions. Thus, the average energy of the gas molecules stays constant unless energy is transferred to or from the gas.

Partial Pressures

One of the interesting properties of an individual gas molecule is that it hits the walls of the container independently from other gas molecules. Thus, two moles of hydrogen will cause the same pressure as one mole of hydrogen and one mole of oxygen. The total pressure of a mixture of gases is the same as the sum of the pressures that each gas would exert individually.

Stoichiometry for Gases

When working with balanced chemical equations, the coefficients of the equation generally refer to the number of moles of the reactants and products.

If we are given a volume of a gas, then we can simply use the P and T of the gas to convert the amount of gas into the number of moles. After we figure out how much product we make, we can calculate the volume of a gaseous product using P and T again. In the special case where the reactants and products are gases and the P and T do not change, we can take the coefficients of the balanced reaction to refer to the volumes of reactant and product.

Example: If we have 300 L of nitrogen gas and 800 L of hydrogen gas at 450°C and 1.15 atmospheres, what volume of ammonia could we make from these reagents?

1.   Balance the reaction: N_2 + 3 H_2 → 2 NH_3.

2.   Determine how much ammonia could be made from each starting material using the coefficients of the balanced chemical equation.

$$300 \text{ L N}_2 \text{ x } \frac{2 \text{ NH}_3}{1 \text{ N}_2} = 600 \text{ L NH}_3,$$

$$800 \text{ L H}_2 \text{ x } \frac{2 \text{ NH}_3}{3 \text{ H}_2} = 533 \text{ L NH}_3.$$

3.   Choose the amount of product that is smaller. This reaction would produce 533 L of ammonia if the yield were 100%.

We could solve this problem by using $PV=nRT$ to calculate the number of moles of hydrogen and nitrogen that we had in the beginning of the calculation, but that would take much longer than the solution that we obtained by working directly with volumes.

Pre-workshop Problems

1.   If you have one mole of a gas at STP, what volume does it occupy?

2. If 25.6 L of a gas at STP is heated to 100°C and its pressure decreases to 0.5 atm, what is the final volume of the gas?

3. Draw a picture of samples of two gases in two different flexible containers (like balloons). One of the gases is at a higher pressure than the other one. Try to illustrate the assumptions of kinetic molecular theory in your illustration.

Cooperative Group Problems

1.   Write Boyle's Law in verbal, graphical, algebraic, and proportional ways. It may help if you calculate the pressure of one mole of gas at 300°C at each of the following volumes:

a. 20 L

b. 40 L

c. 60 L

d. 80 L

e. 100 L

2.   If you have 25.0 L of silicon tetrafluoride gas and 45.0 L of hydrogen gas, what volume of silane (silicon tetrahydride) can you make? Do this calculation without calculating the number of moles of any gas in the system (that will speed the calculation).   What assumption is necessary for this solution to be valid?

3a. You are told that you have a 250 L balloon that is at 25 degrees Celsius and at a pressure of 1.2 atm. What else do you know about the gas in the balloon?

b.   If the temperature of the gas is raised to 300°C and the pressure of the gas increases to 2.9 atmospheres, what is the new volume of the gas?

4.   When you heat a non-flammable gas, what MUST happen to it? What other possible consequences might result from heating the gas? Under what conditions would you observe each possible result? Hint: Think about the kinetic molecular theory. This is a difficult question that tests your logical skills and conceptual understanding of gas behavior.

Practice Exam Questions
Unit 12

1. Boyle's Law can be expressed as

    a. $P/V = k$
    b. $PV = k$
    c. $V/n = k$
    d. $PV = nRT$
    e. $V/T = k$

2. Avogadro's Law can be expressed as

    a. $nV = k$
    b. $PV = k$
    c. $V/n = k$
    d. $PV = nRT$
    e. $V/T = k$

3. A gas in a container is measured at 19.7 L with a pressure of 745 mm Hg. If the container's volume increases to 22.5 L, what is the new pressure?

    a. 851 mm Hg
    b. 652 mm Hg
    c. 0.595 mm Hg
    d. 330 mm Hg
    e. 0.0012 mm Hg

4. Helium was contained in a 30 mL container at a temperature of 25°C. If the temperature of the container increased to 75°C, what size container would be needed to maintain the same pressure?

    a. 346 mL
    b. 26 mL
    c. 35 mL
    d. 90 mL
    e. 10 mL

5. What is the volume of 0.75 moles of an ideal gas at exactly 100°C and 2 atm of pressure?

    a. 3.1 L
    b. 11 L
    c. 14 L
    d. 17 L
    e. 22 L

6. If 5.46 g of bromine gas occupies 15.0 L, what volume will 9.45 g of bromine gas occupy at the same pressure and temperature?

    a. 25.9 L
    b. 0.121 L
    c. 26.9 L
    d. 53.7 L
    e. 12.1 L

7. If 1.00 g of $H_{2(g)}$ and 5.00 g of $O_{2(g)}$ are placed in a 6.00 L container at 45.0°C, what will be the total pressure in the container?

    a. 3.69 atm
    b. 0.403 atm
    c. 26.1 atm
    d. 1.43 atm
    e. 2.85 atm

8. A container holds 4.0 mol of CO_2 and 3.0 mol of N_2. The container has a total pressure of 9.5 atm at 25°C. Calculate the partial pressure of N_2.

    a. 5.4 atm
    b. 4.1 atm
    c. 7.1 atm
    d. 6.3 atm
    e. 3.7 atm

9. In the following equation, 5.77 g of calcium carbonate are heated and carbon dioxide is given off. What is the volume of carbon dioxide given off if the final temperature is 25°C and the final pressure is 0.785 atm?

$$CaCO_{3(s)} \rightarrow CaO_{(s)} + CO_{2(g)}.$$

    a. 1.80 L
    b. 2.44 L
    c. 3.55 L
    d. 7.68 L
    e. 9.90 L

10. How many moles of gas are present in a 5.00 L container at 1.75 atm pressure at 20.0°C?

    a. 5.33 mol
    b. 2.74 mol
    c. 8.41 mol
    d. 0.962 mol
    e. 0.364 mol

11. If 3.00 L of hydrogen reacts with 1.50 L of nitrogen to form ammonia, what is the volume of ammonia that could be produced? The temperature and pressure remain constant during the reaction.

    a. 1.50 L
    b. 2.00 L
    c. 3.00 L
    d. 4.00 L
    e. 4.50 L

12. What is the volume of 2.00 moles of gas at STP?

    a. 2.00 L
    b. 20.0 L
    c. 44.8 L
    d. 48.9 L
    e. 64.0 L

Unit 13: And the Solution Is...

Many times there is some ambiguity in the English language. We are required to decide on the proper meaning of a word by considering the context in which the word is used. You might refer to a stout man or to a glass of stout. The two words are related, but different. The word "solution" could refer to the answer to a question or problem. It could also refer to a homogeneous mixture that is composed of one substance dissolved into another. The aforementioned stout is a solution containing ethanol and other substances dissolved in water. Rather than being a solution to life's problems, overuse of this solution might involve a temporary loss of one's faculties that could create even more problems.

Volumes and Moles

A solution is prepared by dissolving one substance (the solute) in another (the solvent). In the laboratory, it is very simple to measure the volume of a liquid. When we work with a solution, we often use it to deliver a chemical to a container so that a reaction can proceed. The reactant is the solute; thus, we want to know how many moles of the solute we add so we can determine the amount of product we can make. The conversion from a volume of a solution to a number of moles of solute is simple if we have the right conversion factor. That is why chemists prefer to describe the concentration of solutions by using molarity (M), the number of moles of solute per liter of solution. This unit allows for easy conversions between a volume of a solution and the number of moles of solute that it contains:

$$M \ (mol/L) = \frac{\text{moles of solute (mol)}}{\text{volume of solution (L)}}$$

Preparing a Solution

It is essential to understand the techniques used to prepare a solution. To make a 1-liter solution, we do not use 1 liter of solvent. If we have one liter of solvent and add the solute to it, we would end up with more or less than 1 liter of solution. To make a 1-liter solution, we first dissolve the solute in a smaller amount of solvent and then add enough solvent to bring the *final* solution volume to 1 liter.

Calculating Molarity

To calculate the molarity of a solution, we simply divide the number of moles of solute by the volume of the solution in liters.

Calculating Number of Moles

Recall that the purpose of having the molarity of a solution is to enable us to easily convert between solution volume and the number of moles of solute. The volume of a solution in liters multiplied by the concentration in moles per liters gives the number of moles:

Volume (L) x Concentration (mol/L) = Number of moles (mol).

Percent Composition by Mass

Chemists prefer to use molarity as the standard unit to describe the composition of a solution. Many times in industry, solutions are labeled with their composition by mass percent. Percent composition by mass is defined as the fractional mass multiplied by 100. The fractional mass is the mass of the component divided by the total mass.

$$\% \text{ NaOH} = \frac{\text{mass NaOH}}{\text{total mass of solution}} \times 100.$$

Conduction of Electricity

Contrary to popular belief, pure water does not conduct electricity at any reasonable voltages. In order to conduct electricity, water must contain ions in solution. The greater the molarity of ions in solution is, the greater the conductivity of the solution.

Dilution

To avoid shipping many heavy low-concentration reagent bottles, many manufacturers ship chemical compounds in concentrated form. This allows chemists and biologists to easily prepare a solution of any concentration they want by adding more solvent to the concentrate. This process is called dilution. The key to calculations involving dilution is simple: when we add solvent to a solution, we must not change the number of moles of the solute. The number of moles of solute at the beginning of the dilution must equal the number of moles of solute at the end of the dilution. The volume of the solution increases while the concentration of the solution decreases. Using the preceding formula for the number of moles of solute, we have

$$M_i V_i = M_f V_f.$$

Calculations using this equation are a straightforward application of simple algebra.

Stoichiometry

When studying any chemical reaction, we use a balanced chemical equation to calculate how much reactant we need to make a certain amount of product or how much product we can make from the reactants we have. When solutions are involved, stoichiometry calculations are simple, but one additional step is required. We must convert between volumes, concentrations, and numbers of moles.

Titration Reactions

Many people treat titration problems as special, difficult problems in chemistry. Fortunately, they are not. You can treat them like any other stoichiometry problem, except that the amounts of reagents are typically measured and reported differently. Reviewing the steps required to solve stoichiometry problems is useful at this time.

Step 1: Write a balanced chemical equation that describes the reaction of interest.

Step 2: Calculate the number of moles of the reactant or product for which you have a complete set information.

Step 3: Use the stoichiometric coefficients in the balanced chemical equation to calculate the number of moles of the reactant or product of interest.

Step 4: Convert the number of moles to the units the question requires for the answer.

In a titration, a known amount of reactant is added to a solution with an unknown amount of another reactant. The solution will also contain an indicator that changes color when the reaction is exactly completed and both reactants are completely consumed. At the point of color change, a calculation to determine the unknown amount of reactant can be done.

Example Titration Problem

It takes 27.4 mL of 0.1021 M NaOH to neutralize 30.0 mL of sulfuric acid of unknown molarity. Calculate the molarity of the sulfuric acid.

Step 1: $2 \text{ NaOH } + \text{ H}_2\text{SO}_4 \rightarrow \text{ Na}_2\text{SO}_4 + 2 \text{ H}_2\text{O}.$

Step 2: 0.0274 L NaOH x 0.1021 mol/L NaOH = 0.00279754 mol NaOH.

Step 3 0.00279754 mol NaOH   x $\underline{1\ H_2SO_4}$ = 0.00139877 mol H_2SO_4.
                                                   2 NaOH

Step 4 $\underline{0.00139877\ mol\ H_2SO_4}$ = 0.0466 M H_2SO_4.
               0.0300 L H_2SO_4

Pre-workshop Problems

1.   What is the molarity of a 1.5 L solution that contains 15.6 g of sodium hydroxide?

2.   Write a balanced chemical equation for the reaction of hydrochloric acid with calcium hydroxide.

3.   How many moles of phosphoric acid are present in 30 mL of a 0.65 M phosphoric acid solution?

4.   If you dissolve 40.0 g of barium chloride in water and dilute it to a final volume of 500 mL, what is the molarity of chloride in the final solution?

Cooperative Group Problems

1.   You have 500 mL of a 6.00 M hydrochloric acid solution and need to prepare the following solutions for use in a study of reaction rates:

500 mL of 0.10 M hydrochloric acid

500 mL of 0.010 M hydrochloric acid

100 mL of 0.050 M hydrochloric acid

100 mL of 0.0050 M hydrochloric acid

How much of your stock solution will remain after you have prepared the solutions?

2.   You have 300 mL of a solution marked "barium hydroxide."   Before you can dispose of it, you must know the concentration. It takes 54.6 mL of 0.1012 M hydrochloric acid solution to neutralize 30.0 mL of the barium hydroxide solution.   What is the concentration of the barium hydroxide solution?

3.   Which solution will have the highest conductivity?   Remember to write equations to describe the dissociation of each solute (if any).

1 L of solution containing 2.54 g of sodium hydroxide

0.5 L of solution containing 1.30 g of sodium hydroxide

1 L of solution containing 6.7 g of iron (III) hydroxide

0.5 L of solution containing 5.6 g of magnesium chloride

4.   If a bleach solution is 4.3% sodium hypochlorite, what is the molarity of the solution?   The density of the solution is 1.09 g/mL.   Hint: If you assume that you have a 100 g sample of the solution, the problem is much easier to solve. Since concentration is an intensive property, we can assume any amount of solution we want.

Practice Exam Questions
Unit 13

1. By definition, the _____ dissolves the _____ to form a solution.

      a. solvent, solute
      b. solute, solvent
      c. liquid, solid
      d. aqueous solution, salt
      e. water, salt

2. In general terms, the substance that is dissolved to form a solution is the

      a. solvent
      b. solute
      c. salt
      d. aqueous solution
      e. solid

3. Molarity expresses

      a. mass of solute/mass of solution
      b. mass of solute/volume of solution
      c. moles of solute/volume of solvent
      d. moles of solute/moles of solvent
      e. moles of solute/volume of solution

4. When a solution is diluted, what does not change?

      a. the volume of the solution
      b. the concentration of the solvent
      c. the concentration of the solute
      d. the number of moles of the solvent
      e. the number of moles of the solute

5. A solution of aqueous sodium acetate contains 35.0 g of sodium acetate in 45.0 g of water. What is the mass percent of the sodium acetate?

      a. 25.4%
      b. 67.8%
      c. 51.6%
      d. 43.8%
      e. 77.8%

6. A 7.34% calcium chloride solution weighs 13.4 g. What is the mass of the calcium chloride in this solution?

    a. 0.98 g
    b. 0.55 g
    c. 1.83 g
    d. 1.02 g
    e. 0.36 g

7. If 540 mL of 0.435 M HCl is diluted to 1000 L what is the new molarity of the HCl?

    a. 0.806 M
    b. 0.511 M
    c. 0.623 M
    d. 0.235 M
    e. 0.434 M

8. A particular laboratory experiment calls for 150 mL of 0.250 M NaOH. The lab assistant has 500 mL of 0.500 M NaOH. What volume of 0.500 M NaOH and water are needed to make the NaOH solution for the lab?

    a. 150 mL of 0.500 M NaOH, 0 mL of H_2O
    b. 100 mL of 0.500 M NaOH, 50 mL of H_2O
    c. 125 mL of 0.500 M NaOH, 25 mL of H_2O
    d. 75 mL of 0.500 M NaOH, 75 mL of H_2O
    e. 50 mL of 0.500 M NaOH, 100 mL of H_2O

9. A titration experiment is set up in the laboratory. The NaOH that each lab group will use has a molarity of 0.50 M and a volume of 25.0 mL. What volume of 0.1 M HCl will each group need to neutralize this reaction?

    a. 25 mL
    b. 50 mL
    c. 75 mL
    d. 100 mL
    e. 125 mL

10. A certain amount of 1.00 M NaOH will be needed to neutralize 50.0 mL of 0.25 M sulfuric acid. What volume of NaOH will be needed?

    a. 120 mL
    b. 12 mL
    c. 25 mL
    d. 100 mL
    e. 150 mL

Unit 14: Totally Basic

If the pH of your blood were to change by 0.1 units, you would die. Acids and bases can act as catalysts, corrosives, tools for chemical syntheses, or cleaners. Few classes of compounds are so general or important. Eighty billion pounds of sulfuric acid are used every year. That is 80,000,000,000 pounds every year.

Revisiting Definitions

When we discussed the different types of reactions, we encountered a definition for acids that we will continue to use. An acid is any compound that can donate a hydrogen ion (H^+) to another substance. A base is any compound that will accept the hydrogen ion. In this chapter, we will study the reactions of acids and bases in more depth, introduce more terminology, and quantify acid strength.

Conjugate Acids and Bases

Under the right circumstances, many reactions can be reversed. If a compound releases a proton (H^+), the product of the reaction might accept the proton back in the future, re-forming the reactant:

$$HI + NaCl \rightarrow NaI + HCl \text{ (HI is the acid, Cl}^- \text{ is the base).}$$

The reverse of that reaction is

$$NaI + HCl \rightarrow HI + NaCl \text{ (HCl is the acid, I}^- \text{ is the base).}$$

Also, when a compound that is a weak base accepts a proton, it might lose it later to a strong base:

$$HI + NaCl \rightarrow NaI + HCl \text{ (HI is the strong acid, Cl}^- \text{ is the weak base);}$$
$$NaOH + HCl \rightarrow H_2O + NaCl \text{ (HCl is the acid, OH}^- \text{ is the base).}$$

In each of the preceding examples, the product of the reaction of a base with an acid later acted as an acid itself. Specifically, after Cl^- accepted a proton to form HCl, the HCl could potentially act as an acid. The Cl^- is called the base, while the HCl is the conjugate acid that is formed after the base does its thing. Likewise, HI was the acid in the first reaction and I^- was the conjugate base, which has the potential to accept the H^+ back. If you need to know the conjugate base of any acid, just remove the H^+ and see what is left.

Complete Dissociation

Strong acids are those acids that dissociate fully in water to give ions. We can write the reaction of a strong acid dissolving in water as

$$HCl(g) \xrightarrow{H_2O} H^+(aq) + Cl^-(aq).$$

However, it is more accurate to write

$$HCl(g) + H_2O(l) \xrightarrow{H_2O} H_3O^+(aq) + Cl^-(aq).$$

The hydrogen ion does not sit by itself in the water; it bonds to a water molecule, forming the hydronium ion, H_3O^+. Water acts as a base in this instance, accepting a proton.

The Strong Acids

You currently do not have a good way to predict which acids are strong and fully dissociate in water, so you should memorize some of these strong acids. The six common strong acids are hydrochloric acid, hydrobromic acid, hydroiodic acid, nitric acid, sulfuric acid, and perchloric acid. When a strong acid is placed in water, the solution conducts electricity very well, as there are many ions formed in the solution.

The Weak Acids

In the nomenclature chapter, we learned to name quite a large number of acids. All the acids except the preceding six strong acids listed are weak. This means that most of the acid molecules fail to donate a H^+ to water. The acid stays primarily in its undissociated form.

Water Reacts With Itself

Water can act as a base when it reacts with acids to form H_3O^+, which is commonly abbreviated as H^+. Water can act as an acid when it reacts with bases to form OH^-. Thus, water can act as either an acid or a base. From this, it follows that water can react with itself to form ions:

$$H_2O(l) + H_2O(l) \rightarrow H_3O^+(aq) + OH^-(aq).$$

This reaction only proceeds to the extent that the concentration of H^+ ions will be 1×10^{-7} M and the concentration of OH^- ions will be 1×10^{-7} M in pure water.

The Ion-Product Constant

If you multiply the concentration of hydronium ions in water by the concentration of hydroxide ions, the resulting value is 1×10^{-14} M^2. Remarkably, the product of multiplying the concentration of hydronium ions by the concentration of hydroxide ions has the same value in any aqueous solution. If an acid releases more H^+ into solution, some of the H^+ will react with OH^-, reducing both concentrations until the ion-product constant has a value of 1×10^{-14}:

$$[H^+][OH^-] = 1 \times 10^{-14} M^2 = K_w,$$

Here $[X]$ means the concentration of X and K_w is the ion-product constant for water.

The pH scale

Hydrogen ion concentrations in aqueous solution can range from 10^{-14} M to above 1 M. Since humans vastly prefer to work with whole numbers, scientists have decided to describe hydronium and hydroxide concentrations in aqueous solution using a logarithm-based system:

$$pH = -\log [H^+].$$

Since $\log 10^{-14}$ is -14, then $-\log 10^{-14}$ is 14. Since the general range of $[H^+]$ is 10^{-14} to 1, pH will range from 14 to 0. The solution with the lowest pH is the one with the highest $[H^+]$.

Buffers

A solution can be "buffered" to prevent the pH of a solution from changing if acids or bases are added to it. If we mix a weak acid and a weak base together in the same solution, the solution will be protected from the effects of strong acids or bases. Any strong acid that is added to the solution will react with the weak base, removing the added H^+ from the solution. Likewise, any strong base that is added to the solution will react with the weak acid component of the buffer, neutralizing the strong base. A mixture of acetic acid and sodium acetate is one such buffer.

Pre-workshop Problems

1.   Write the balanced chemical equations for each of the following reactions.

a. Acetic acid with a strong base.

b. Sodium acetate with a strong acid.

2.   For each acid or base in the following table, identify whether it is an acid or base. For each acid, write the formula of its conjugate base. For each base, write the formula of its conjugate acid.

Compound	Conjugate
HNO_3	
NH_3	
CO_3^{2-}	
HCl	
OH^-	
$HC_2H_3O_2$	

3.   Fill in the table with the missing quantities and decide if the solution is acidic or basic. Each line of the table represents the acidity or basicity of a different solution.

$[H^+]$	$[OH^-]$	pH	pOH	Acidic or basic?
0.012 M				
	1×10^{-12} M			
		7.96		
			2.1	

Cooperative Group Problems

1.	Pretend that this sheet of paper is a representation of a dilute aqueous solution containing equal amounts of hydrochloric acid and formic acid. If you were able to see molecules and ions, draw what you would see. Make sure you consider what atoms would really look like and the relative concentrations of all species in solution.

2.   The strong acids have very weak conjugate bases, and the very weak acids have strong conjugate bases. Have each person in your group offer their own explanation using an analogy, if possible. After everyone is done, bring the group back together and decide which explanation is the most clear.

3.   Someone has prepared a buffer solution for you by adding 40 g of NaOH to one liter of a 2 M phosphoric acid solution.

a.   What are the concentrations of all species in solution that you have received?   Hint: Always start with a balanced chemical equation.

b.   What happens to the buffer solution if one mole of perchloric acid is added to it?

Practice Exam Questions
Unit 14

1. Acids produce _____ in aqueous solutions.

    a. hydrogen
    b. hydroxide ions
    c. water
    d. hydronium ions
    e. none of the above

2. Bases produce _____ in aqueous solutions.

    a. hydrogen
    b. hydroxide ions
    c. water
    d. hydronium ions
    e. none of the above

3. In the following reaction, identify the conjugate acid of NH_3.

$$NH_3 + H_2O \rightarrow NH_4^+ + OH^-.$$

    a. H_2O
    b. H^+
    c. OH^-
    d. NH_4^+
    e. none of the above

4. Which is the correct conjugate base for CH_3COOH?

    a. CH_2COOH^-
    b. CH_3COO^-
    c. OH^-
    d. $CH_3COOH_2^+$
    e. none of the above

5. The correct formula that defines pH is

    a. $\log[H^+]$
    b. $\log[OH^-]$
    c. $-\log[H^+]$
    d. $-\log[OH^-]$
    e. none of the above

6. Which equation can be used to calculate pOH?

    a. $pOH = log[OH^-]$
    b. $pOH = 14 - pH$
    c. $pOH = -pH$
    d. $pOH = -log[H^+]$
    e. $pOH = -log[OH]$

7. The hydroxide ion concentration of a solution has been measured as 3.56×10^{-7} M. Calculate the pH of this solution.

    a. 6.45
    b. 7.55
    c. 3.56
    d. 4.33
    e. 2.45

8. What is the concentration of the hydrogen ion in a solution with a pH of 12.4?

    a. 12.4 M
    b. 1.60 M
    c. 3.98 M
    d. 1.09×10^{-13} M
    e. 3.98×10^{-13} M